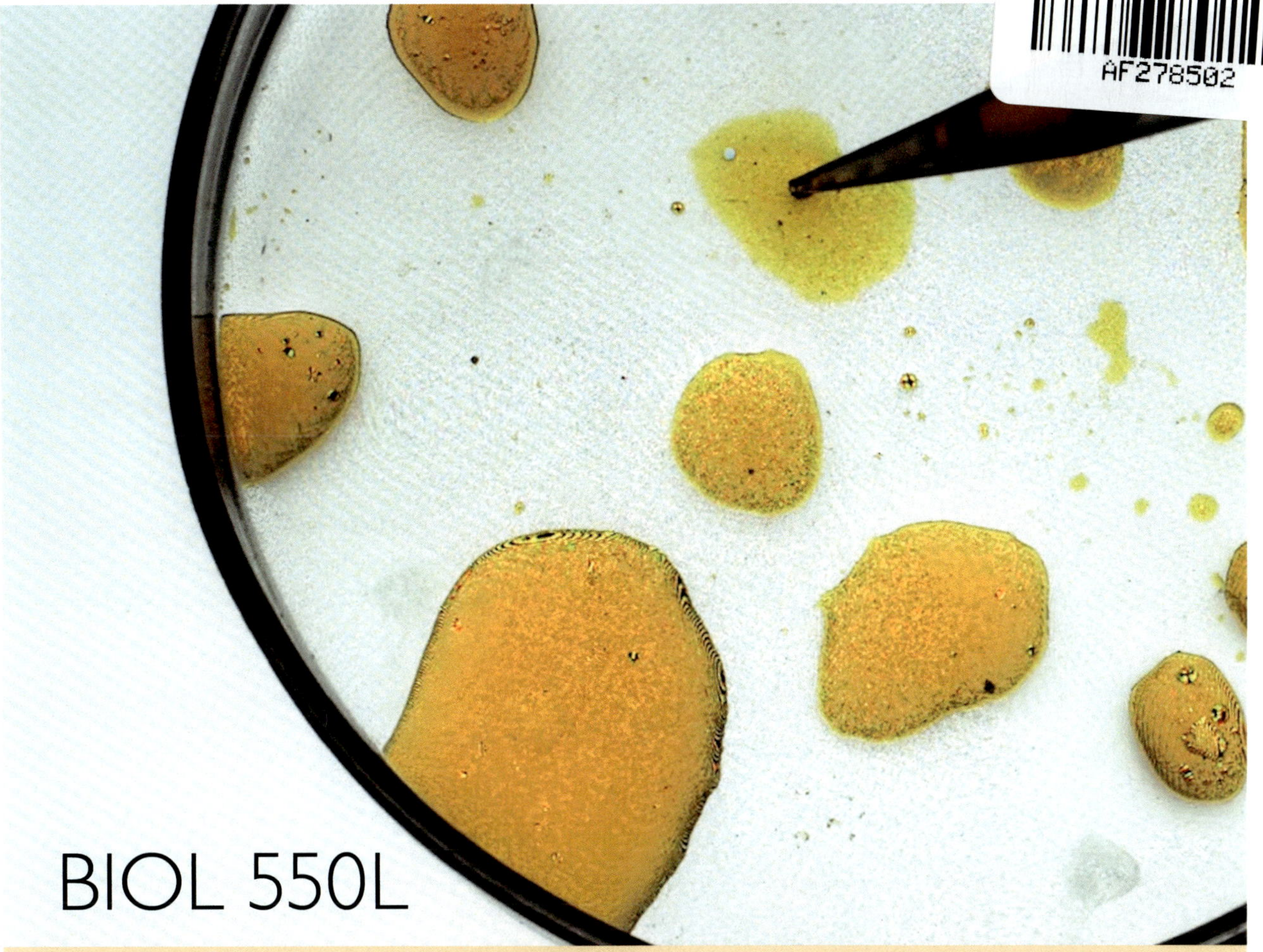

BIOL 550L

BACTERIOLOGY LABORATORY

UNIVERSITY OF SOUTH CAROLINA, COLUMBIA

DEPARTMENT OF BIOLOGICAL SCIENCES

SECOND EDITION

Isaac M. Hagenbuch, Phd
Contributing Author Louis Berrios

VAN-GRINER LEARNING

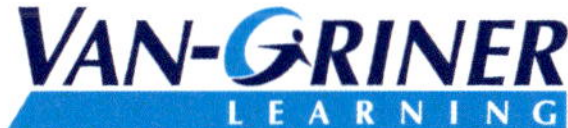

Bacteriology Lab

BIOL 550L
Second Edition
University of South Carolina, Columbia, Department of Biological Sciences
Isaac M. Hagenbuch, PhD
Contributing Author: Louis Berrios

Printed in the United States of America
10 9 8 7 6 5 4 3 2 1
ISBN: 978-1-61740-691-1

Van-Griner Learning
Cincinnati, Ohio
www.van-griner.com

President: Dreis Van Landuyt
Project Manager: Maria Walterbusch
Customer Care Lead: Lauren Houseworth

Hagenbuch 691-1 F19
316216
Copyright © 2021

TABLE OF CONTENTS

Factors That Influence Microbial Growth

Assignments

PREFACE
MICROBIOLOGY LABORATORY RULES

VIOLATION OF THESE RULES WILL RESULT IN FORFEITURE OF POINTS

1. Attendance is required as mandated by the University of South Carolina policy. *All* lab sessions (classes and return sessions) are to be attended by *all* students. If you are late to lab, you cannot take the quiz for that class, and there will be no make-ups. Absence due to a contagious illness should be documented by a medical professional, and any concession is at the discretion of the lab instructor.

2. **Cheating, which includes but is not limited to plagiarism and falsification, will not be tolerated.** *All* work must be your own (even if data are collected as a group) — all lab reports must be your own thoughts and written in your own words. Do not copy another student's work, and do not copy out of the lab manual. There is a zero tolerance policy in the Department of Biological Sciences towards plagiarism. ***Plagiarism = Failure. All*** incidents of cheating will be reported to the Office of Academic Integrity.

USC STATEMENT OF ACADEMIC INTEGRITY

Academic ethical behavior is essential for an institution dedicated to the promotion of knowledge and learning. The University of South Carolina is committed to fostering a university environment which exemplifies the values embodied in the *Carolinian Creed*. All members of the University Community have a responsibility to uphold and maintain the highest standards of integrity in study, research, instruction, and evaluation as well as adhering to the *Honor Code*.

UNIVERSITY OF SOUTH CAROLINA HONOR CODE

The *Honor Code* is a set of principles established by the university to promote honesty and integrity in all aspects of a student's academic career. It is the responsibility of every student at the University of South Carolina to adhere steadfastly to truthfulness and to avoid dishonesty in connection with any academic program. A student who violates, or assists another in violating the *Honor Code,* will be subject to university sanctions.

I understand that attendance is required and that any failure to abide by the *Carolinian Creed* and *Honor Code* will result in a loss of points.

Name: ____   Date: ____________________

RULES REGARDING SAFE & APPROPRIATE LABORATORY CONDUCT

1. You will at some points in this lab be working with potential human pathogens. Therefore, consider all microorganisms you work with in this lab to be potential pathogens and handle them safely. The Microbiology Laboratory rules are for your protection and the protection of people *outside* the lab to whom you may carry microbial pathogens.

2. No smoking/vaping, eating, drinking water bottles, or snack foods in the lab. If you need a drink or a snack, wash your hands and partake of it outside of the lab. Do not chew gum, the ends of pens, your fingernails, etc. Your mouth is a wide-open portal of entry.

3. Glass pipettes require the use of a pipetter. *No mouth pipetting* is permitted.

4. Always wear shoes with closed toes. If you do not, you will be asked to leave the lab and change your shoes to ones with closed toes.

5. Long hair must be tied back because we work with flames.

6. **Wash your hands thoroughly (10–15 sec) with soap and warm water before starting work and before leaving the laboratory (even if you are just visiting the restroom).**

7. *Never under any circumstances* take cultures out of the laboratory.

8. **Notify the instructor** *immediately* **in the case of personal accidents —** *no matter how small.* These include cuts, burns, spills, any introduction of microbial material into the eyes, mouth, or broken skin, etc.

9. Swab the bench top with 5% Lysol at the beginning and end of each laboratory period. If a culture is spilled, cover the spill with a Lysol-soaked paper towel and notify the instructor.

10. Long pants are required in lab to cover and protect your exposed lower half. Natural-fiber clothing you don't mind permanently staining is an acceptable substitute. The stains used to stain cells in this laboratory are *permanent dyes* and cannot be fully removed from clothing.

11. Gloves should be used when handling live cultures or stains. Skin is generally a sufficient barrier for all microbes we use in this lab, but gloves provide an additional measure of safety to both you and those outside of the lab.

12. Use the shelves at the front of the lab to store coats, hats, backpacks, books, etc. All such things, except your lab notebook and manual, must be kept off of the bench-top. Your possessions can become contaminated by simple contact. Anything upon which a culture is spilled must be autoclaved to rid it of living microbes.

13. ***Do not use the sinks for waste disposal*** unless otherwise directed by your instructor. This includes disposal of cultures.

14. Worn gloves, used solid media, disposable inoculation instruments, and culture spill clean-up materials must be discarded into the labeled biohazard containers. **Under no circumstances will you discard these items in the standard trash cans.**

15. All other items such as paper towels, parafilm wrappers, scraps of paper, inkless pens, exhausted pencils, etc. will be discarded in the black trash bins located at the end of each bench and by the exit.

16. When collecting media for the day's exercises, take only the number of tubes and plates specified. Identify each medium by marking with a Sharpie.

17. **All plates and tubes must be properly disposed of in the specified containers after data are recorded.** Test tubes are to be labeled with your Sharpie and the labels must be removed (using ethanol) from ***all*** test tubes before they are put in racks for sterilizing.

18. Glass pipets and micropipette tips must be placed in designated containers after use.

19. Never open an agar plate that shows signs of fungal growth. Keep the fungal spores contained.

20. **Students should wear safety glasses when working with a flame.**

21. Keep alcohol away from any flame. You will be required to work with alcohol and flames during the course of lab. Caution is to be used whenever you are working with a flame but ***especially*** when alcohol is involved.

22. Bunsen burner flames must be kept low. Never let them exceed 3 inches in height.

23. Microscopes are numbered and will be inspected. They must be returned ***clean and to their proper space.*** All oil must be removed from the objectives using lens paper.

24. The syllabus is a contract, which can only be modified by the TA or Lab Coordinator. Any modifications will only occur with sufficient notice.

25. Expect that labs will always take the full allotted time — **don't ask if we'll end early.**

26. Turn cell phones to silent or vibrate. **Cell-phone disturbances will *not* be tolerated.** Text/tweet/snapchat/facebook/tinder at your own risk. Failure to pay attention in lab can get you hurt and will get you a failing grade.

27. **Late work policy:** Lose half a letter grade (5%) per day late. **e.g.:** B+ would be dropped to a B.

28. If you have a question about anything, ask the TA.

29. It is **your** responsibility to prepare for labs and **your** responsibility to listen to directions in lab.

30. Clean up before you leave. Don't leave a mess or a disorganized space for those who come after you.

31. Class disturbance is unacceptable. If you are a distraction to others or present behavior problems, you will first be asked to desist. If you continue, you *will* be asked to leave the lab. If you refuse, USC police will be called to remove you.

32. If you arrive to lab short of sleep and show evidence of this, you may be asked to leave. Tired people are a hazard to themselves and others, especially in a lab with fire and infectious organisms.

I HAVE READ RULES 1–32 AND AGREE TO ABIDE BY THEM WHILE IN LAB.

Name: ____ Date: ____________________

ASEPTIC TECHNIQUE & CULTURE TRANSFER

Microorganisms intended for laboratory and scientific investigation are grown and kept in **pure cultures.** It is a regrettable reality that microorganisms are not visible to the unaided human eye. One consequence of this is that technicians have no immediately visible way of determining if they are keeping their microbial cultures pure. Therefore, the first technique that any student of microbiology must learn is **aseptic technique.**

Aseptic technique is a set of best practices and procedures designed to

- facilitate the creation of pure microbial cultures;
- keep microbial cultures pure and uncontaminated;
- keep the microbes only in certain, specific places and at specific times; and
- keep people outside the lab safe by not transferring any microbes from the lab into the larger world.

Every microbial culture we use is many times more concentrated than what you would find in the natural world. This means that they all present a certain amount of danger to those who handle them. There are so many microbial cells present that exposing cuts, mouths, ears, eyes, etc. directly to them can result in an infection that will require medical treatment. Properly executed aseptic technique and adherence to the laboratory safety rules will ensure that you reduce your risk level to the lowest level possible.

GOVERNING PRINCIPLES OF ASEPTIC TECHNIQUE

- **Fire as a sterilizing tool.** Inoculation loops and needles will be sterilized by heating until they glow orange in a Bunsen burner flame (Figures 1 and 2). Once sterile, loops/needles must be cooled sufficiently before contacting any microorganisms to avoid killing them. Glass rods will be stored in ethanol and residual ethanol burned off before the rod is cooled and used. Glass cools slower than metal.

- ⊙ **Use of adductive vertical air currents.** When working with open nutrient media, whether inoculated or not, work within the envelope of adductive (moving toward center) and vertical (upward) air currents created by the Bunsen burner flame (Figure 3). This will keep dust particles and their attendant microorganisms from settling into your plates. Lightly flame the necks of test tubes before re-capping to create a convection current that will carry air *out* of the tube.

- ⊙ **Limiting exposure of nutrient media.** Media must be capped or lidded at all times unless you are transferring organisms or material in or out of the container. Tubes of medium should be held at an angle and not vertically. Caps and lids must not be placed on the benchtops.

- ⊙ **Keeping silent and attentive while inoculating media.** Accidents happen when you are inattentive, therefore keep your attention on your work. Also, do not talk while making organism transfers as saliva-containing aerosols coming out of your mouth can contaminate your medium or your transfer instrument. Do not blow onto inoculating instruments to cool them!

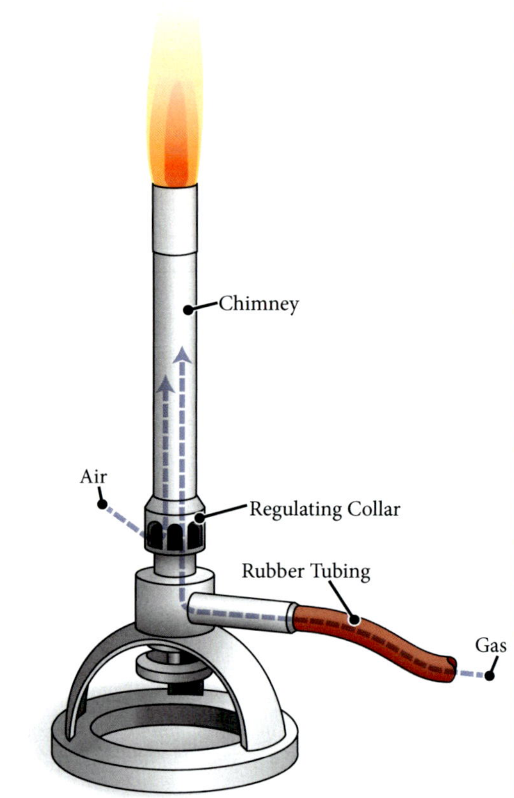

- ⊙ **Avoiding the creation of bacteria-containing aerosols.** Always cool your inoculation, transfer, or spreading instrument to a temperature that won't kill the bacterium nor boil the medium in which they reside. Boiling, shaking, sonication, and other agitation methods can produce bacteria-containing aerosols that can lead to contamination not only of your nutrient medium but of your lab bench, skin, clothing, or lungs.

- ⊙ **Storing and incubating solid medium plates inverted.** Always incubate and store your solid media plates in an inverted position to prevent any condensation that may appear from flowing onto the medium surface. Dust sticks to water droplets. Therefore, any of this sort of water/medium contact may contaminate your media either before or after inoculation.

FIGURE 1

Bunsen Burner. A Bunsen burner is a simple natural gas jet. Its main parts are the metal chimney and the air intake regulator. The chimney allows the gas and air to mix before burning at the top of the chimney. More air = hotter flame, less air = cooler flame. Balance the air and gas to get the type of flame you need.

⊙ **Disinfecting work surfaces.** Surfaces, such as table tops, should be disinfected with 70% ethanol (EtOH) or Lysol before and after each use. This serves to keep the number of potentially contaminating microorganisms to a minimum. Even having done this, do not set down things like inoculation loops, lids, or caps unless it is absolutely essential (such as to deal with a safety issue). If you must place a cap on a benchtop, always place it with its inner surface facing downward.

PROCEDURE

1. Collect an **Eppendorf tube** containing live bacterial culture.

2. Obtain two additional tubes of sterile **tryptic soy broth (TSB)** and label them (A & B, 1 & 2, whatever convention you want to use).

3. Light your Bunsen burner and adjust it until you have a well-defined consistent flame (Figures 1 and 2).

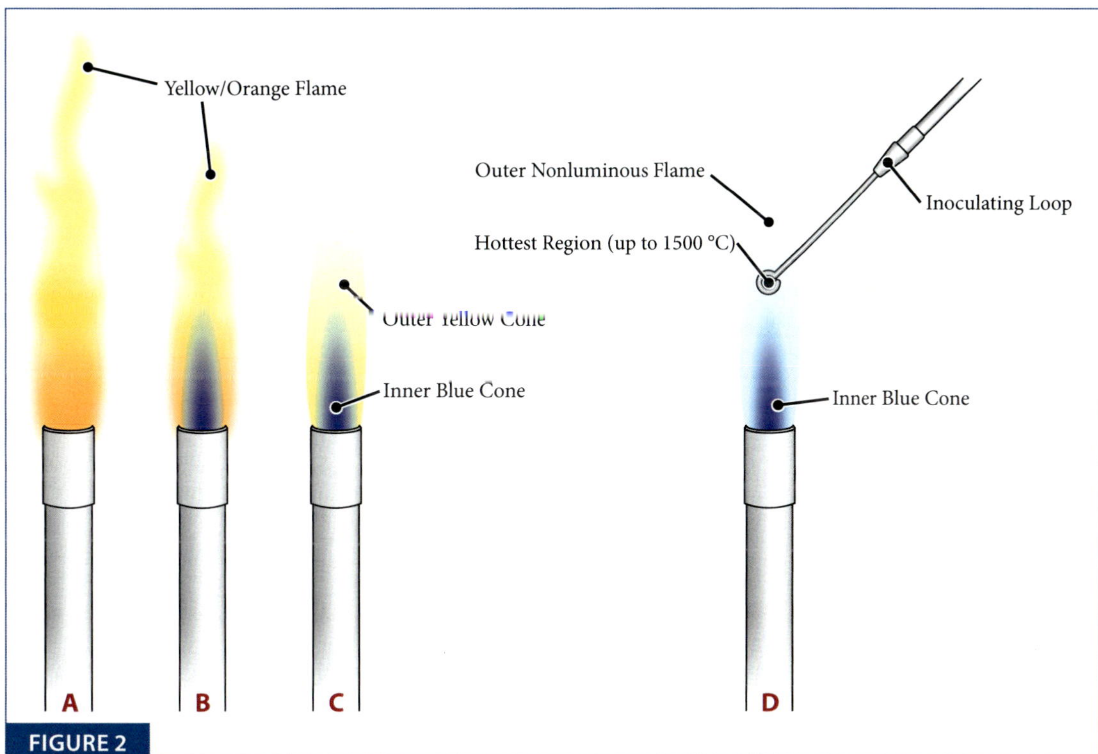

FIGURE 2

Bunsen Burner Flames. Here are some examples of the types of flames you get with different gas/air mixtures. The general rule is that a yellow/orange flame is burning inefficiently (not hot enough). A) Is a result of too little air and is not useful for sterilizing because it is too cool. B) Is closer to the ideal flame but there is still too little air/too much gas. C) Is useful for sterilizing but not yet the most efficient. D) Is the ideal where there is no indication of inefficient burning and there is a bright blue flame, indicating the hottest possible fire.

4. Sterilize your inoculating loop within the flame by heating the wire until it glows orange-red.

5. Remove your loop from the flame and cool it (count to 10). Do not blow on it! Your breath can contain droplets of saliva that carry bacteria which will contaminate your inoculation instrument.

6. Use your inoculating loop to collect a sample from the live culture and transfer it to one of your tubes of TSB.

7. Flame your loop again to sterilize it. Wire must glow orange-red to be sure of sterilization!

8. Cool the loop and place it in the *other* tube of TSB.

9. Remove the loop and flame it again before putting it aside.

10. Incubate the TSB tubes for 24 hrs at 37 °C.

11. After incubation, you should have a live bacterial culture (cloudy) in your first tube, and the second tube should still be sterile (clear).

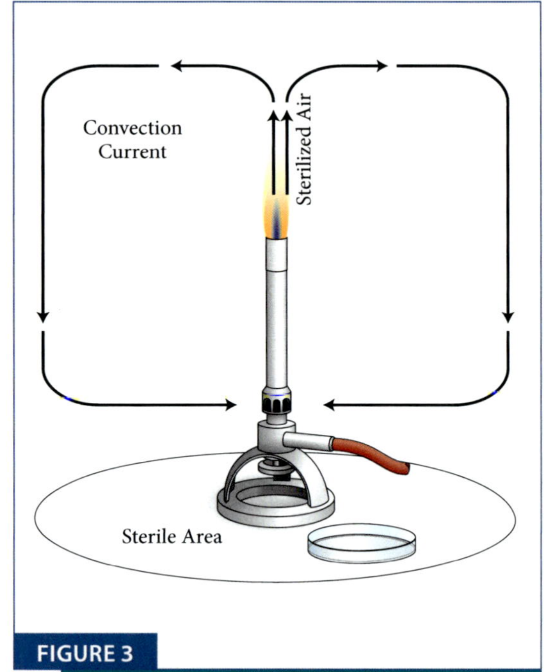

FIGURE 3

Sterile Area. Hot air rises and draws the air below it upward. On a tabletop, this results in a 3D donut of convection currents surrounding a Bunsen burner flame. The lateral and upward movement of the air will direct dust particles and aerosols away from your medium surface.

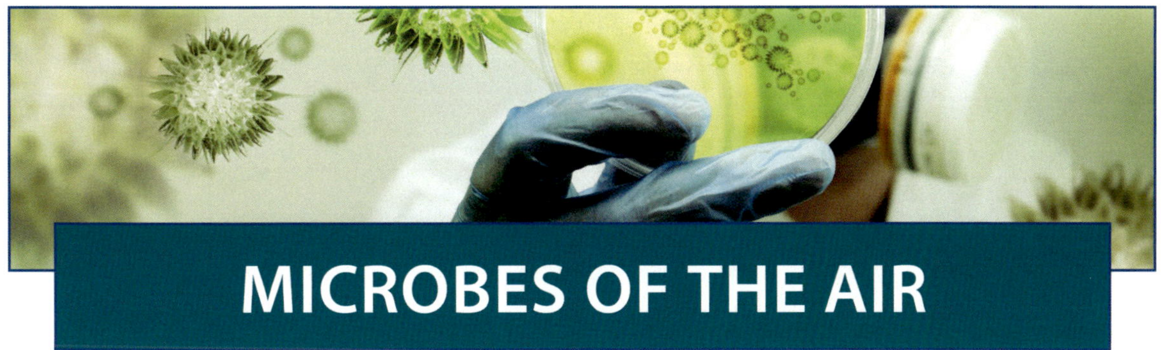

MICROBES OF THE AIR

Every type of microbe may be found in the air at various times and in various places. Airborne organisms are transient. There is no known microorganism that carries out any major part of their lifecycle preferentially while airborne. Within building interiors fungi are generally the most numerous **culturable microorganism** in the air followed by bacteria.

We are able to culture ~1% of the microbes that we have genomic evidence for. Clearly, we have much work yet to do in this area. Though we have had limited success so far, the creation and maintenance of microbial cultures is a key skill in both scientific and clinical pursuits. Many factors contribute to the successful culturing of any microorganism, including the following:

- Culture medium
- Temperature
- Moisture
- Gasses

You should therefore not expect that any medium will allow you to culture *all* the microorganisms that are present in a given environment. In this experiment, you will be using **tryptic soy agar (TSA)** to determine if there are culturable microbes in the air of the laboratory. TSA may allow the growth of some fungi, but it will also allow the growth of a wider array of bacteria, if present.

PROCEDURE

1. Obtain a TSA plate and label it with your name and section.
2. Open the lid of the plate and leave the medium exposed for some period of time (note how long).
3. Close the lid and **parafilm** it (see Figure 1).
4. Incubate at room temperature for up to one week.

5. If, after incubation, the plate has anything fuzzy growing in it, ***do not open it.*** That fuzziness is likely a fungus that is rife with fungal spores. If you open it, you risk spreading spores hither and yon.

6. Choose a bacterial colony to describe and describe it using proper cultural-characteristics terms.

FIGURE 1

How to Parafilm Your Petri Dish. Parafilm will adhere to itself once stretched out. Separate the Parafilm from its backing and stretch it completely around your Petri dish and firmly attach the end. Be careful that you cover the entire gap between the lid and base. Inspect your Parafilm job for holes before placing it in the incubator. Holes in your parafilm can result in a dried-out, ruined culture.

SUBCULTURING IN BROTH & ON SOLID MEDIA

The transfer of live, pure microbial cultures from one place to another and without contamination is an indispensable skill for anyone who may need to collect or preserve microbial specimens in a professional/clinical context. When transferring microbes, intentional transfer is called **inoculation** and inadvertent transfer is called **contamination.** Generally, we want to maximize our ability to inoculate while simultaneously minimizing contamination. Toward this end, you will employ aseptic technique as you perform inoculations to and from liquid and solid microbiological media.

When streaking on a solid medium, the goal is to obtain well-separated (discrete) colonies. The streaking technique is a type of dilution. You are diluting the culture by spreading it over a greater and greater surface area. You want to spread the culture out enough that you are able to separate individual cells far enough apart that they can give rise to separate colonies composed of millions of cells just like the original cell. These are what we call pure colonies, and we must be able to consistently obtain them so we can be sure that we are maintaining the purity of our microbial cultures.

PROCEDURE

1. Obtain for yourself:
    a. One tube of bacterial culture.
    b. One Petri dish (plate) of sterile tryptic soy agar (TSA).
2. Being careful to observe aseptic technique:
    a. Sterilize your inoculation loop.
    b. Transfer live bacteria from the culture onto the plate using a standard streaking technique.
    c. Sterilize your inoculation loop.
    d. Transfer live bacteria to the TSA plate while successfully completing a standard streaking technique and obtaining a goodly amount of discrete colonies.

Your laboratory instructor will demonstrate the proper technique, and you may also refer to Figure 1.

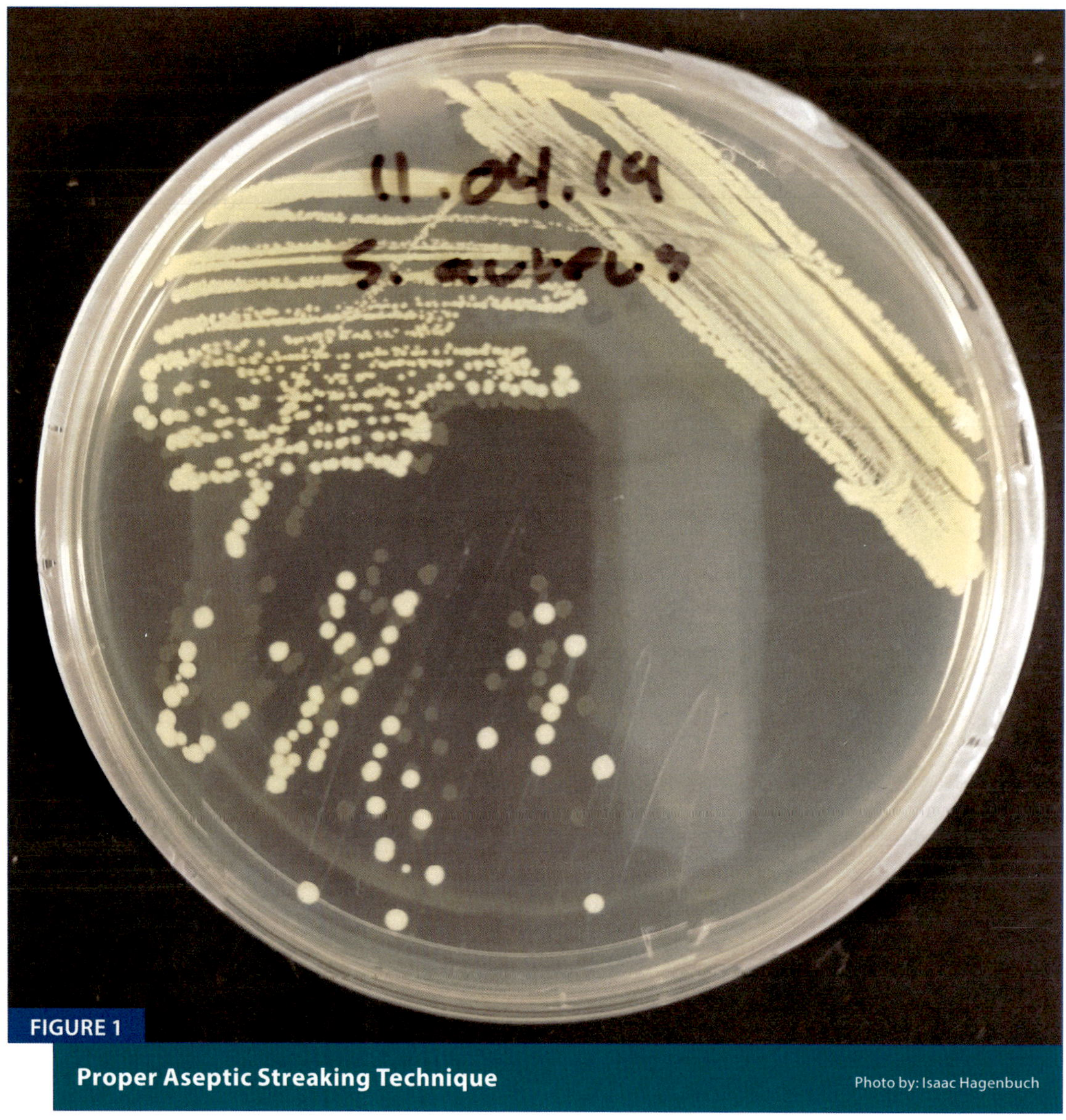

FIGURE 1

Proper Aseptic Streaking Technique

Photo by: Isaac Hagenbuch

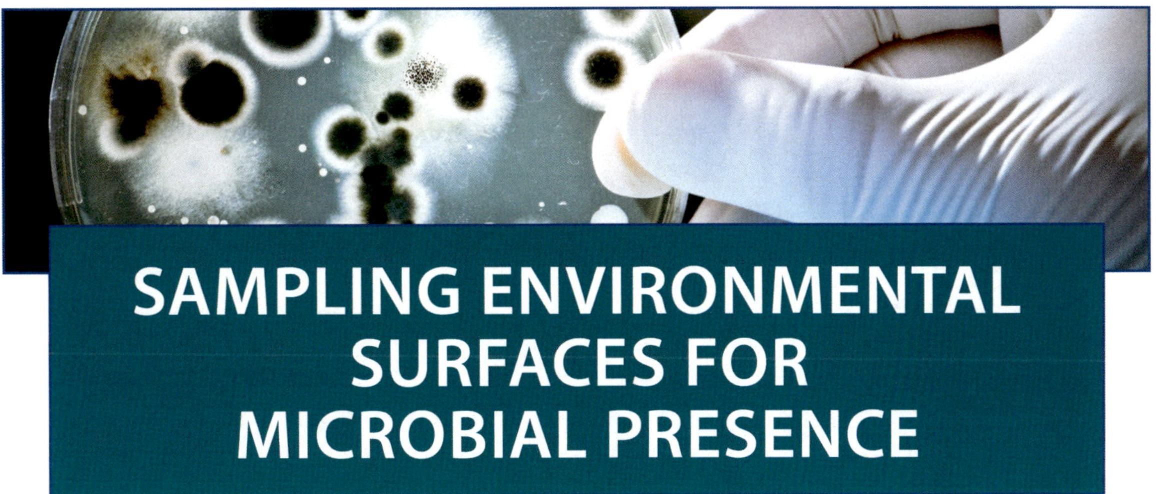

SAMPLING ENVIRONMENTAL SURFACES FOR MICROBIAL PRESENCE

You may have been told that microbes are everywhere, but do you know that for certain? Are the same number and types of microorganisms present on any given environmental surface? These are questions you will begin to answer by the end of this laboratory exercise.

The same caveats that apply to culturing microbes from the air also apply here. There is no way to culture *all* microorganisms that may be present on environmental surfaces. Your results will therefore only allow you to say something about whether there are culturable microorganisms present.

If for no other reason, then at least for the sake of your own self-respect, don't choose some basic location to sample. What is a basic location? Anything within arms-reach as you sit in your chair. The lab benchtop, drawer handles, the inner handle of the microlab door, the lab faucet handles, etc. Take your sampling materials and leave the lab room. Go find someplace that may actually be interesting. Floors, walls, and windows are generally cold and don't have enough direct human contact to carry anything.

PROCEDURE

1. Obtain a tube of sterile saline, a sterile swab, and a TSA plate.

2. Find a non-basic surface to test.

3. Wet the swab with the saline. Recap but keep it in your hand.

4. Take your moist swab and rub it all over your chosen sampling location. Be sure to press firmly and to rotate the swab as you sample to get an accurate representation of the organisms present.

5. Pop the saline cap again and submerge the swab in the saline. Close the lid on the swab stick to brace it and then *break the handle off* leaving the swab inside the container.

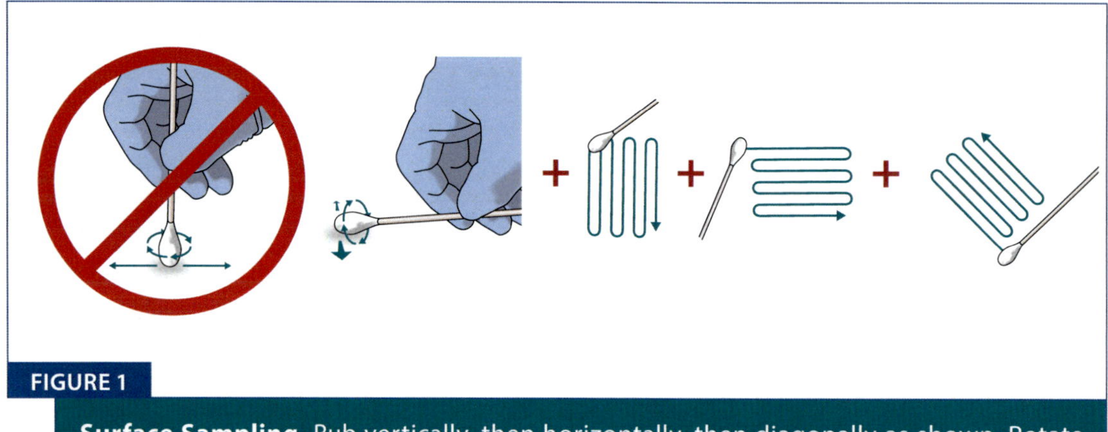

FIGURE 1

Surface Sampling. Rub vertically, then horizontally, then diagonally as shown. Rotate the swab whilst rubbing.

6. Return to lab and find a free vortexer. Vortex your swab container for 10–20 seconds. Make sure the liquid is actually spinning within the container.

7. Bring your swab back to your station and use your pipette to create a spread plate with 250 µl of the liquid.

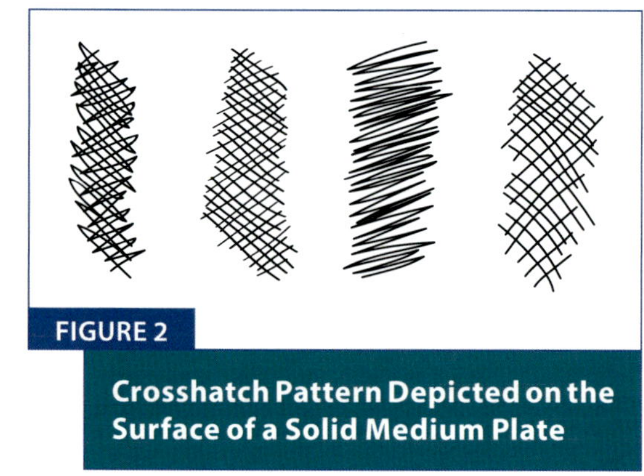

FIGURE 2

Crosshatch Pattern Depicted on the Surface of a Solid Medium Plate

   a. Remove a glass spreader from the ethanol beaker and light the alcohol on fire with the Bunsen burner and let the alcohol burn off. ***Do not hold the glass in the Bunsen burner flame.***

   b. Let the spreader cool (count to 30 or more).

   c. Place your plate on a turntable or lab bench, open the lid (do not set it down!), place the glass spreader flat against the medium surface, and rotate the plate as you use the spreader to spread an even layer of the sample over the medium surface.

   d. Replace the lid and parafilm it shut.

8. Discard the swab container into the regular trash.

9. Incubate the plate at room temperature for up to one week.

10. After incubation, if you find anything fuzzy growing in the plate, ***do not open it.***

11. Observe a bacterial colony and describe its colony characteristics.

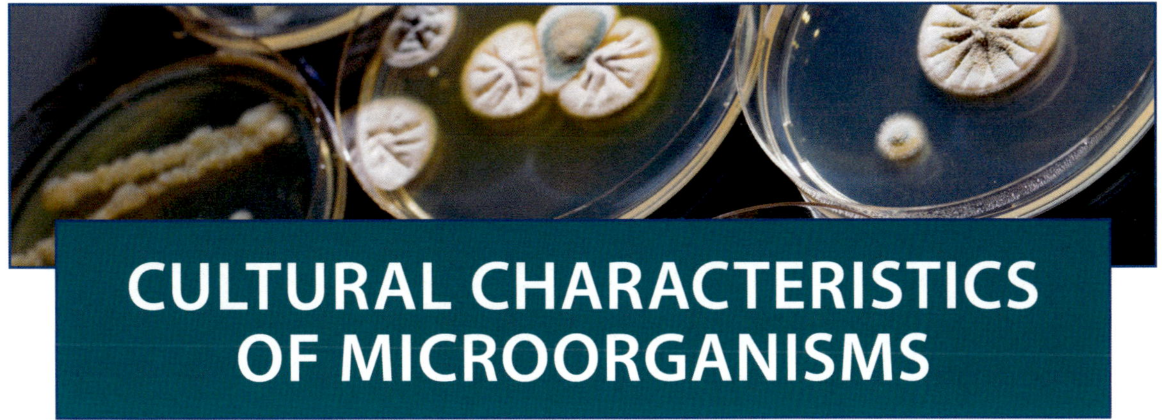

CULTURAL CHARACTERISTICS OF MICROORGANISMS

The appearance of microbial cultures is often useful in identification of microbes. Differences in appearance are manifested in a number of characteristics that can vary according to species, but it is important to keep your culture conditions constant. Why? Because cultural conditions (temperature, nutrient profile, medium additives) can cause a single species to show variation. The studious person can find cultural characteristics for most known bacterial species within the volumes of *Bergey's Manual of Systematic Bacteriology*, but a basic knowledge is sufficient to be useful in most contexts.

This laboratory exercise will give you the terms you need to use when describing colonies from your air exposure plate, environmental swab plate, and in various exercises through the rest of the semester.

NOTE: If any of your plates have filamentous (fuzzy) colonies, ***do not open the plate.*** These colonies are likely fungi, and opening the plate may release fungal spores into the air of lab. Fungal spores are already a common contamination issue, so let us not make the issue worse.

Choose a colony in your air and surface plate and apply some combination of the following **cultural characteristic** terms to a description of them.

CULTURAL CHARACTERISTIC TERMS

NUTRIENT AGAR PLATES

This is perhaps the most commonly used type of microbiological medium. It is very useful for maintaining the purity of cultures. These are often inoculated using 3- or 4-way streak technique. Discrete colonies on solid media may differ in the following ways:

1. **Size** – Pinpoint, small, medium, large.
2. **Pigmentation** – Colony color. Some organisms are **chromogenic** and will therefore produce pigments that color the colony or the medium upon which it grows. Most organisms are non-chromogenic and therefore appear a shade of white, gray, or tan.

3. **Form** – The shape of the colony which may be one of four basic types.

    a. **Circular** – A smooth, unbroken periphery.

    b. **Irregular** – An edge with a non-uniform indentation.

    c. **Rhizoid** – Growth that spreads in a root-like way.

    d. **Spindle** – Ovoid, or eye-shaped growth with a swelled middle and pinched ends.

4. **Margin** – The appearance of the outer, leading edge of the colony.

    a. **Entire** – Even (not bumpy), sharply defined.

    b. **Lobate** – Marked, irregular indentations.

    c. **Undulate** – A wavy appearance.

    d. **Erose** – Serrate. Regular, sawtooth-like.

    e. **Filamentous** – Fuzzy, thread-like.

5. **Elevation** – The difference in height between the surface of the medium and the colony surface.

    a. **Flat** – No discernable difference in elevation.

    b. **Raised** – Colony surface slightly above medium surface.

    c. **Convex** – Hemispherical or dome-like.

    d. **Pulvinate** – Convex but with a higher dome elevation and a resulting larger volume.

    e. **Umbonate** – A two-step elevation consisting of a raised portion with a convex portion on top of it.

The same organism will generally yield additional information if grown in/on other types of media. While we don't use these as often in the lab, it is worth having a familiarity of the terms associated with them.

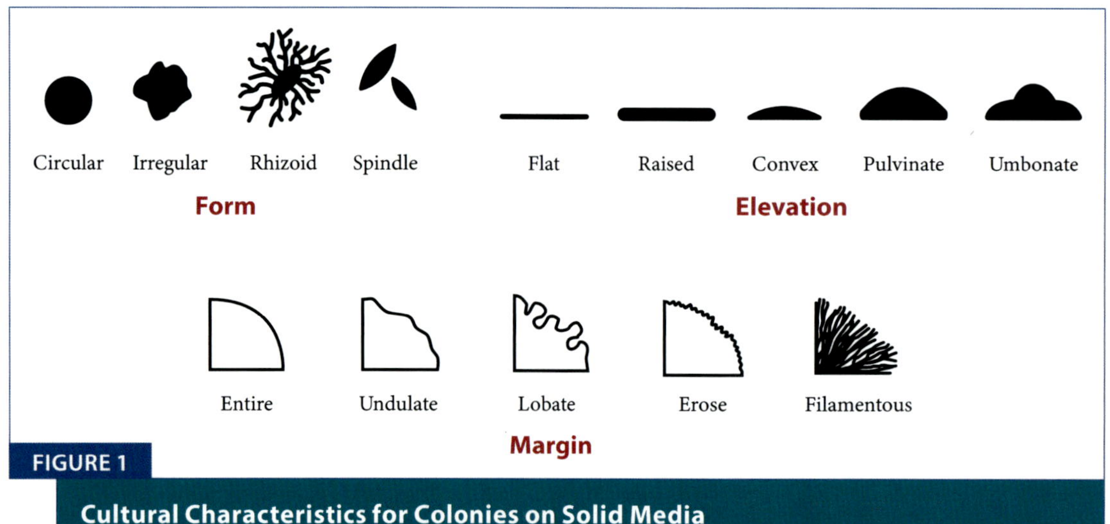

FIGURE 1

Cultural Characteristics for Colonies on Solid Media

NUTRIENT BROTH

Here, you are looking for the distribution and clarity of growth.

1. **Uniform/fine turbidity** – A non-particulate, finely-dispersed growth throughout the medium.

2. **Flocculent** – Flaky or dust-bunny-like aggregates throughout the medium.

3. **Pellicle** – A firm, pad-like growth covering the surface.

4. **Sediment** – Most of the visible cells are congregated at the bottom of the culture vessel. Often flocculent or grainy.

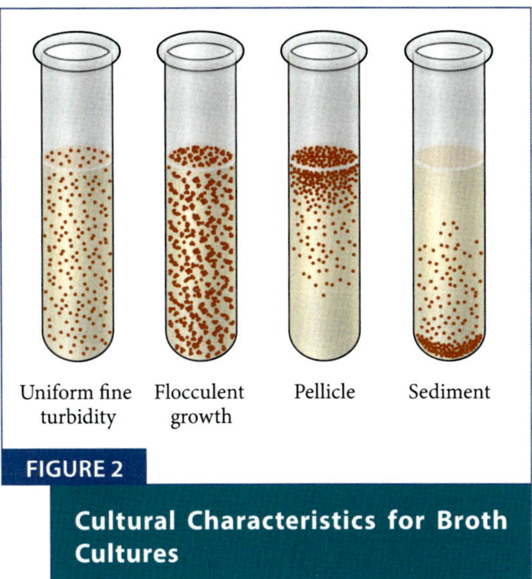

FIGURE 2

Cultural Characteristics for Broth Cultures

NUTRIENT AGAR SLANT

These use the same medium you find in nutrient agar plates, but in a test tube set at an angle as the liquid agar cools. These are inoculated with a single, straight line.

1. **Abundance of growth** – None, slight, moderate, heavy.

2. **Pigmentation** – Color of the colonies.

3. **Optical properties** – Determined by how much light travels through the colony growth: Transparent (all light makes it through), translucent (some diffuse light makes it through), or opaque (no light makes it through).

4. **Form** – The appearance of that single-streak inoculation.

    a. **Filiform** – Continuous growth with smooth edges.

    b. **Echinulate** – Continuous growth with irregular edges.

    c. **Beaded** – Non-continuous, semi-continuous growth.

    d. **Effuse** – Thin, spreading growth.

    e. **Arborescent** – Treelike.

    f. **Rhizoid** – Rootlike.

5. **Consistency** – This has to do with how firm or moist the colonies are.

    a. **Dry** – Apparently free of moisture.

    b. **Buttery** – Moist and shiny like room-temperature butter.

    c. **Mucoid** – Glistening, shiny, slimy.

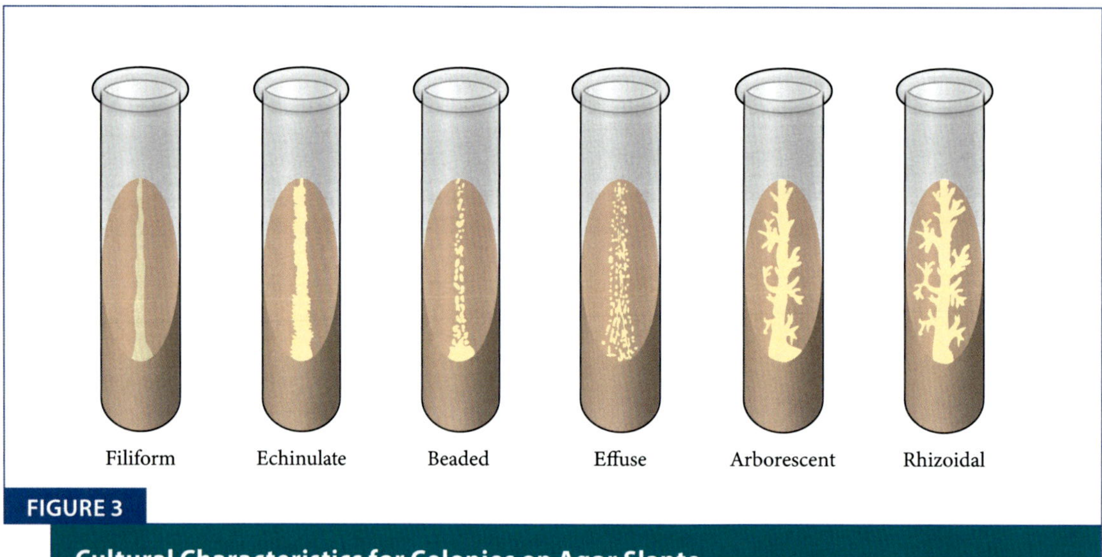

FIGURE 3

Cultural Characteristics for Colonies on Agar Slants

NUTRIENT GELATIN

This medium is commonly used when identifying unknown organisms because the ability to liquefy hardened gelatin is not common to all organisms. This medium is inoculated by stabbing once. Take care to be consistent because the liquefaction pattern is influenced by depth of inoculation.

1. **Crateriform** – Saucer-shaped liquefied surface
2. **Napiform** – A bulb–shaped liquefaction at/just under the service.
3. **Infundibuliform** – Funnel-shaped.
4. **Saccate** – Tubular, extending down through the medium.
5. **Stratiform** – Complete liquefaction of the top half of the medium.

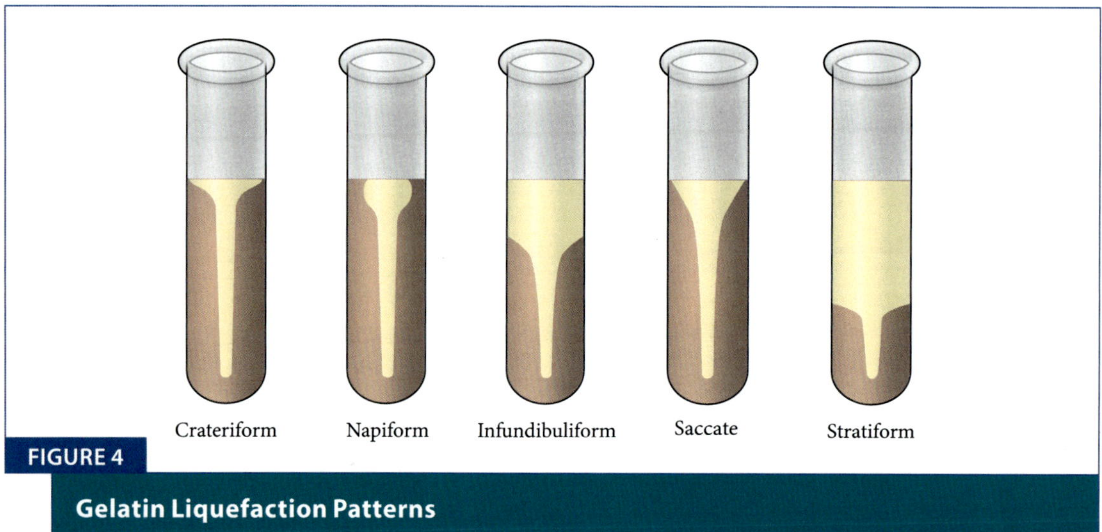

FIGURE 4

Gelatin Liquefaction Patterns

SEPARATING A MIXED CULTURE

In nature, microbes don't self-segregate into uniform populations. Rather, they are found as mixtures of many cell types. This is true in what you might consider "environmental" samples, but it is also true in clinical situations. Wound infections, especially in chronic sufferers, are commonly caused by more than one type of bacteria. Both environmental microbiology and clinical practice depend on being able to create pure cultures; therefore, methods for separating a mixed bacterial sample are foundational techniques.

In this lab exercise, you will use a standard streaking technique upon a solid medium to separate out a mixed bacterial culture. The culture is a mixture of 10% *Escherichia coli* to 90% *Staphylococcus aureus* — a 1:9 ratio. After incubation, you will observe your streak plate and use your skills in identifying cultural characteristics to assess your ability to isolate pure colonies of both species and also note which organism seems to be most abundant.

FIGURE 1

Colonies of *E.coli*, *Staphylococcus aureus*, *Enterococcus faecalis* and *Corynebacterium striatum* on tryptic soy agar (TSA). Cultivation 24 hours, 37 °C in an aerobic atmosphere.

By Hans Newman. Used with permission. https://microbiologyinpictures.com/bacteria-photos/escherichia-coli-photos/bacterial-colonies-on-agar.html

PROCEDURE

1. Obtain a tube of the mixed culture and two TSA plates.

STREAK PLATE

2. Observing proper aseptic technique, transfer a loop of the mixed culture to the TSA plate and perform a standard streak.

SPREAD PLATE

3. Place the second TSA plate upon the turntable.

4. Observing proper aseptic technique, use your pipette to transfer 100 µl of the mixed culture to the center of the plate.

5. Remove a glass spreader from its ethanol bath and use the Bunsen burner to burn the alcohol off. **Do not** hold the glass in the flame; simply use the flame to light the alcohol then remove it from the flame. Leaving the spreader in the flame will overheat the spreader and can kill the culture as you spread it over the plate.

6. Allow the spreader to cool for at least 30 seconds, then use it to spread the mixed culture over the entire surface of the plate.

7. When finished, place the spreader back into its ethanol bath.

8. Incubate at 37° C for 24 hours.

9. After incubation, observe the plate.

    a. Can you discern two different colony morphologies?

    b. Note whether you managed to obtain pure colonies of either species.

    - *E. coli* is tan, irregular, undulate, and raised.

    - *S. aureus* is a lighter tan or yellow, circular, entire, and convex.

    c. Which species is more abundant? Does that comport with what you started with? Why?

    d. Which technique was better suited for separating and observing the colonies and their morphologies?

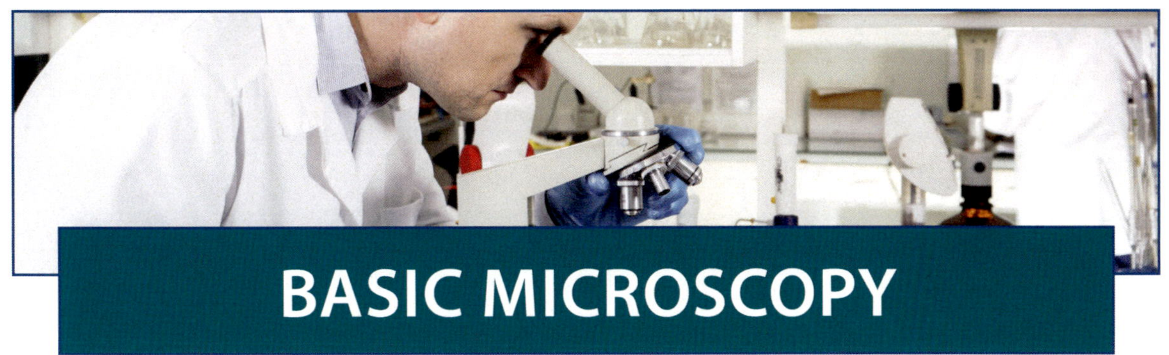

Microscopes are technology and microscopy is a discipline. Both are an outgrowth of a very old area of study: optics. The term optics comes from the (transliterated) Greek *optikē* which means "appearance" or "look." Optics began as a study into how human sight works, but it has become much more than that over the millennia.

From the study of optics first came eyeglasses (sometime around 1289 AD), then spyglasses (first attempt at patenting was in 1608), followed by telescopes (work which also started in 1608), and finally, around 1665, it led to the subject of this chapter: microscopes. Robert Hooke (England) and Anton van Leeuwenhoek (Netherlands) were contemporaries who separately developed two types of microscope. Leeuwenhoek's instrument used a single biconvex (spherical) lens, and Hooke's used *two* convex lenses. It is in Hooke's design, the compound microscope, that all modern microscopes find their origin.

All modern microscopes are compound in design, which is to say they use more than one lens. In this laboratory, we use some of the simplest and least expensive modern scopes: brightfield microscopes (Figure 1). These are part of an extended family of tools, all of which have their own benefits and liabilities.

MODERN MICROSCOPES

BRIGHTFIELD

These microscopes may have two lenses but are more commonly found with five or more, most of which are housed in the **objective.** In this type of instrument, specimens are illuminated from below via a light source that is itself focused by another lens below the stage called a **condenser.** This results in specimens that appear dark against a light background. In smaller organisms, like live, unstained bacteria, this lighting arrangement can be an issue. A brightfield system has limited contrast, which means that the differences that can be drawn between specimens and their surrounding medium are small. This is a problem when observing live organisms. Most observations of bacteria are therefore made on nonviable, stained preparations, and under oil immersion (Figure 2) because of their small size (Figure 3).

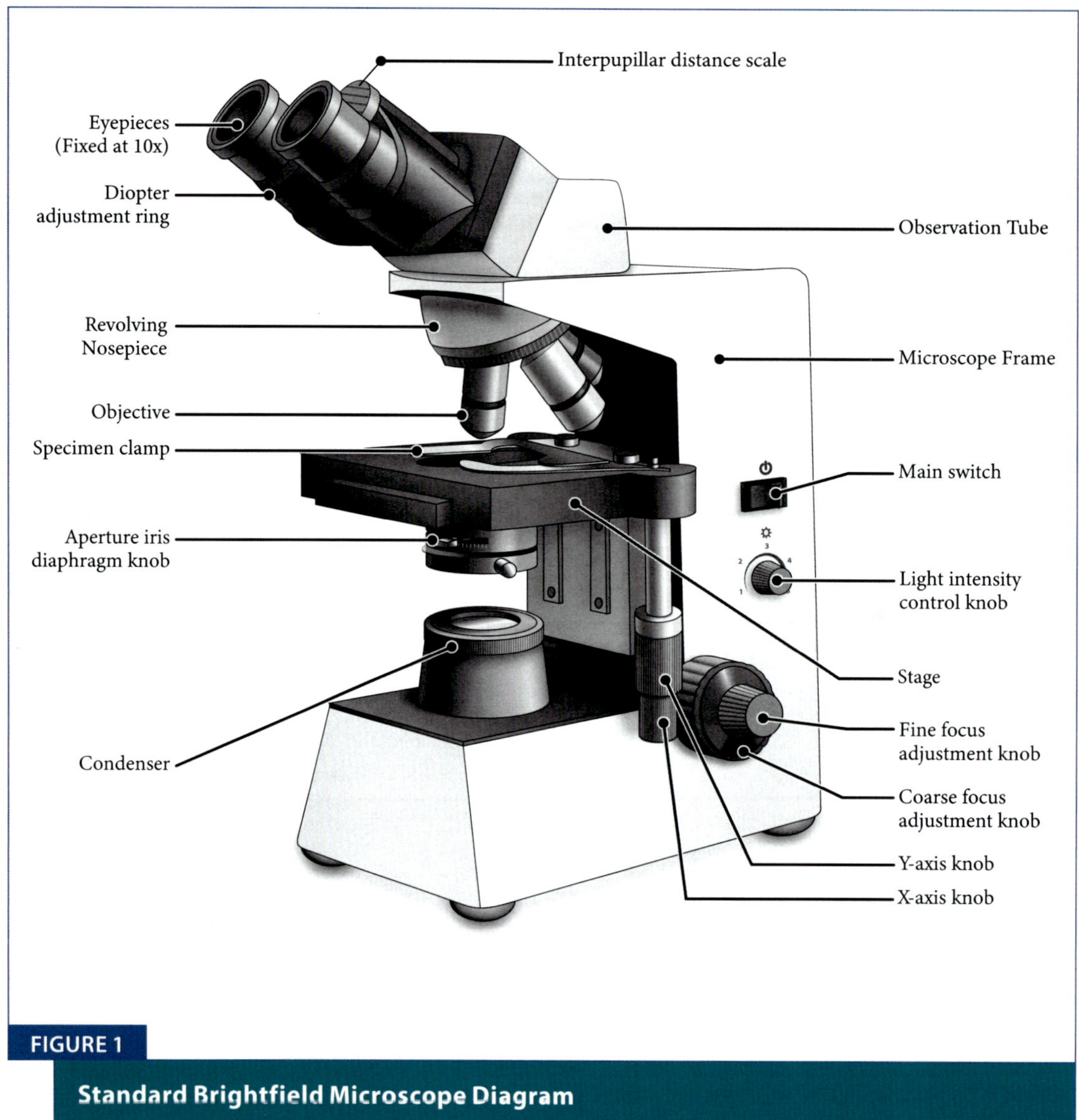

FIGURE 1

Standard Brightfield Microscope Diagram

DARKFIELD

These microscopes have an ocular lens system essentially identical to brightfield micro-scopes. The main difference is in the design of the illumination/condenser system. In darkfield scopes, light is directed upon the specimen from the side so that light is reflected off of it rather than transmitted through it. This yields a light specimen on a dark background. These instruments are able to create greater contrast between speci-mens and their surrounding medium, so they are much more useful for observing both larger, opaque specimens and live microbial specimens.

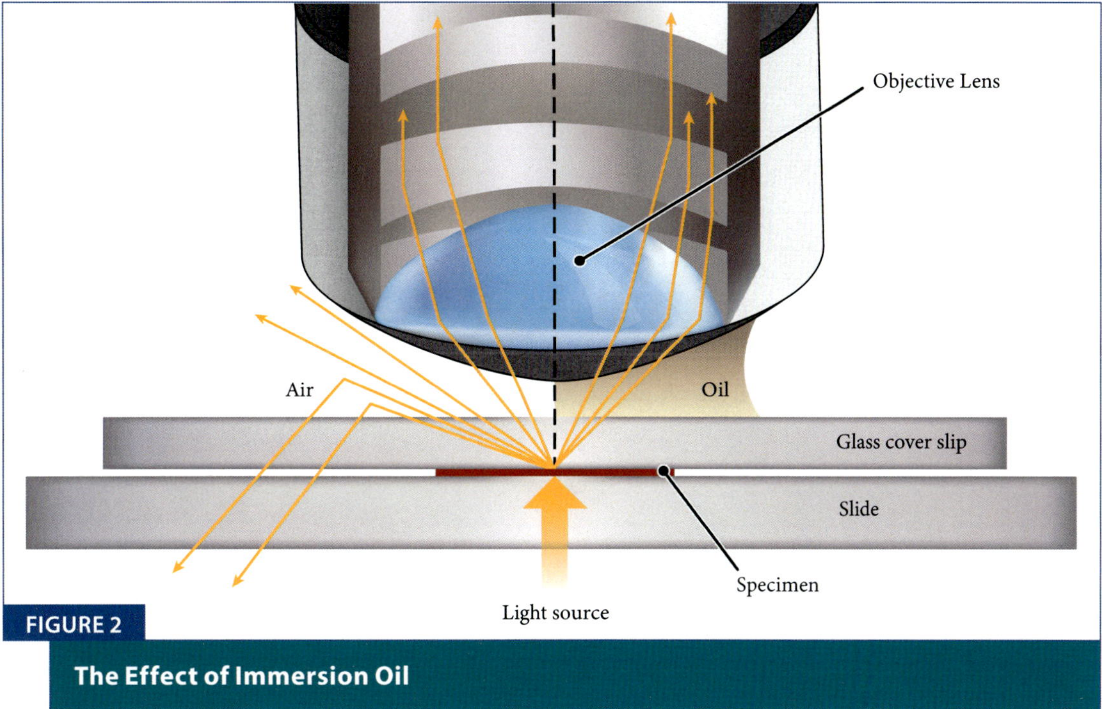

FIGURE 2

The Effect of Immersion Oil

PHASE-CONTRAST

These microscopes include specialized objectives and condensers that accentuate subtle differences in the refractive indexes of cellular components. Amplifying these differences raises the contrast of various intracellular components. This is particularly suited for observing live, unstained microorganisms. These specimens appear dark on a light background.

FLUORESCENT

These microscopes can generally function like standard brightfield microscopes, but they also have some wider uses. They have specialized light sources which are designed to cause fluorescent dyes to glow. These scopes also include optical filters that will block out the excitation light while allowing the fluorescent light to be observed, making it easier to detect your chosen response. These scopes are used in conjunction with any of a large number of fluorescent dyes wherein the microbial specimens are stained with the dye, then observed for fluorescence. This is particularly useful in observing specimens containing very small cells or contaminating objects (like soil particles) that would otherwise make delineating organism from non-living particle impossible.

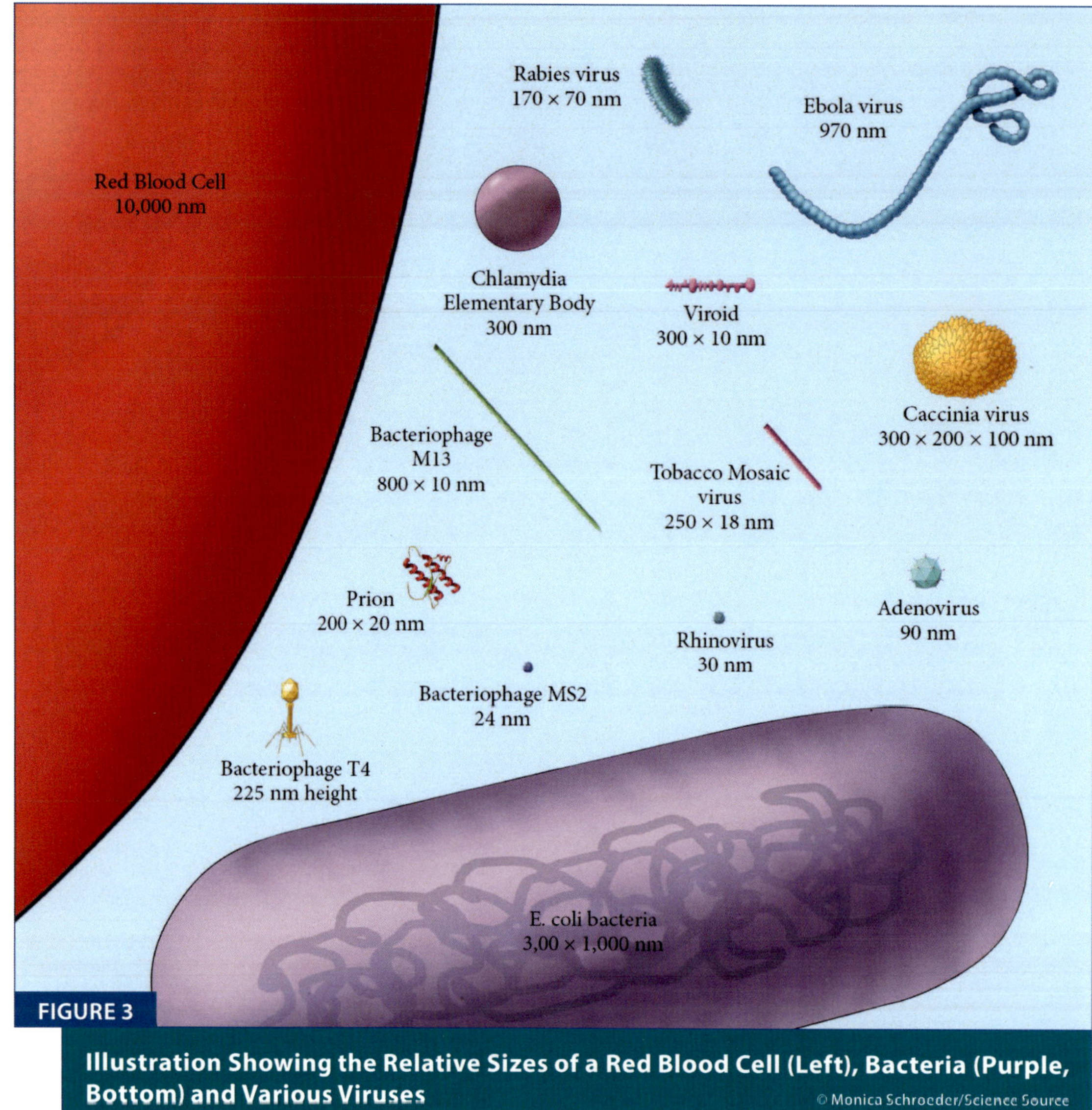

FIGURE 3

Illustration Showing the Relative Sizes of a Red Blood Cell (Left), Bacteria (Purple, Bottom) and Various Viruses

© Monica Schroeder/Science Source

ELECTRON

These microscopes use beams of electrons and not visible light to create images of their specimens. They also do not use traditional optics, since traditional optics would absorb or scatter all of the electrons. Therefore, the optics used here are composed of electromagnetic fields, and the distance from the electron source and the specimen must be put under vacuum because even air would disturb the beam. There is more than one type of electron microscope, but they all work by differentiating between areas of different electron absorbance or reflectance.

PREPARATION OF MICROBIOLOGICAL SMEARS

Specimens of microorganisms must be prepared in some way if they are to be visualized by a microscope. The preparation procedure is determined by the organism in question, the type of microscope you will be using, and what you want to see. Such procedures might include suspending them in liquid media, using special dyes that fluoresce when struck by light of a certain wavelength, or freezing the specimen in liquid nitrogen. In this lab, the basis for most of our microscopic observations will be the **bacterial smear.**

A bacterial smear is exactly what it sounds like: a sample of a bacterial culture or colony smeared onto a microscope slide. Such preparations are decidedly low-tech, but they are the backbone of microbiological observations.

PROCEDURE

1. Obtain a new microscope slide, handling only the sides so as to avoid transferring oil from your fingers onto the work surface.

2. Light up a Bunsen burner.

3. Remove the packaging oil from the slide by passing the top and bottom surfaces briefly through the flame. If the slide has other soiling on it, you can wash the slide using soap and water but be sure to thoroughly dry the slide before proceeding.

4. Label one of the short edges of the slide with a marker pen. Be sure your label is one that will allow you to determine whether you have the slide right side up.

5. Obtain a bacterial culture(s).

6. Take some amount of the culture and smear it onto the slide such that you have a smear 2–3 cm wide.

   a. If you are working with liquid cultures, use a transfer loop and transfer no fewer than 5 loopfuls of the liquid to the slide; you may need to use more than 5 depending on the organism and the concentration of cells.

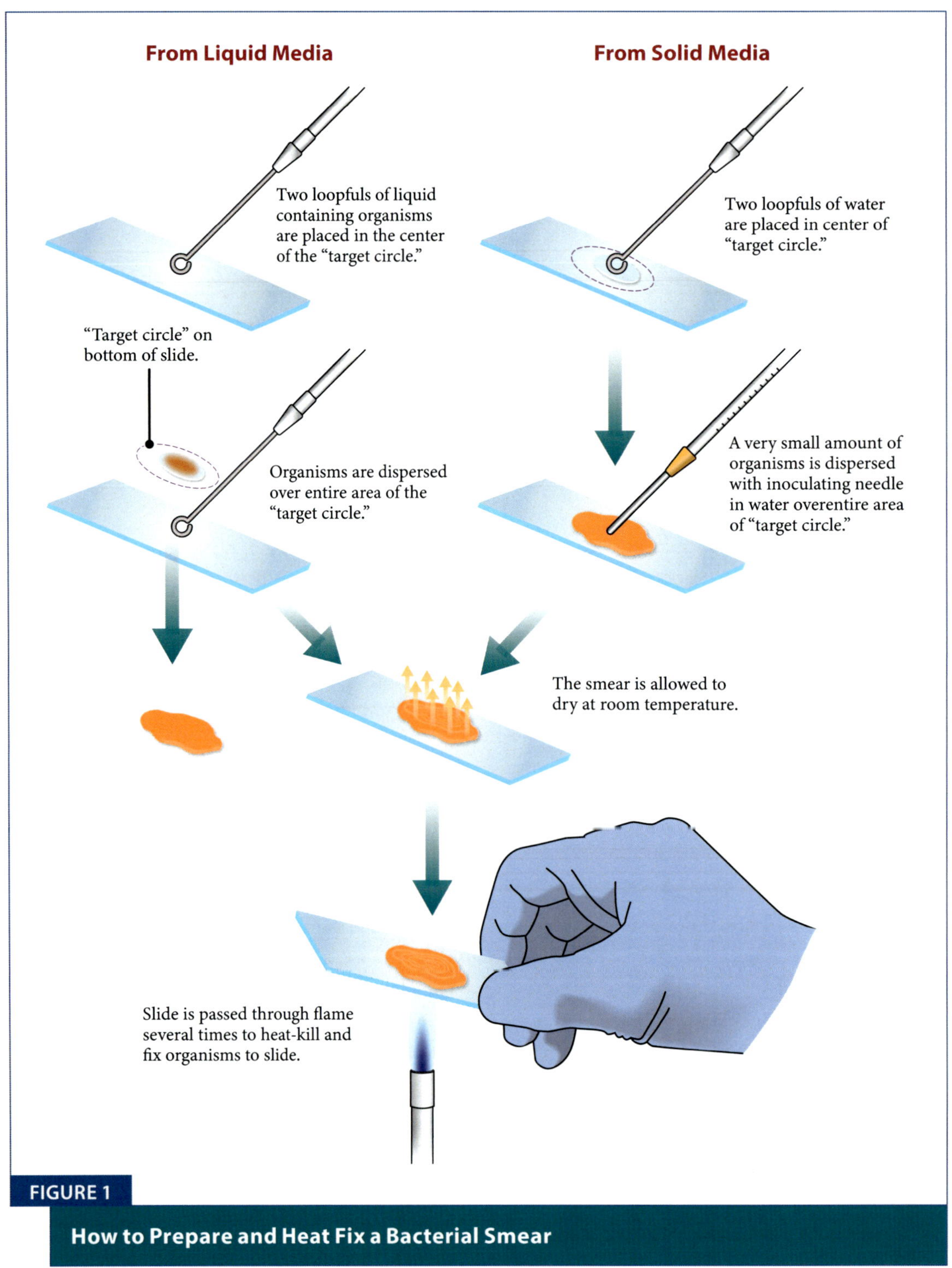

FIGURE 1

How to Prepare and Heat Fix a Bacterial Smear

b. If you are working with colonies on agar, add 20 µl of water to the slide prior to transferring the colony sample. Use the water to dilute and homogenize your smear. Too much water will make a mess, and no water will probably yield a smear that is too concentrated to properly observe.

7. Allow the smear to dry completely.

    a. You can let it air dry on the staining rack.

    b. If you want to speed it up a bit, you can *lightly* heat the bottom of the slide to encourage the water to evaporate quicker. If you heat it too much, you will cause the bacterial cells to lyse, and you will therefore see nothing when it comes time to observe them.

    c. ***Do not blow*** on the slide or wave it around. This encourages contamination of the sample.

8. The smear should be semitransparent. If it is translucent or opaque, it is too thick, and you'll need to make another.

9. Once the smear is completely dry, **heat fix** the smear to the slide by heating the underside of the slide in the flame. It should get hot enough that you could burn yourself if you touched the wrong end but not so hot that you crack the slide or char the smear.

10. Set smear aside to cool before moving onto the next exercise.

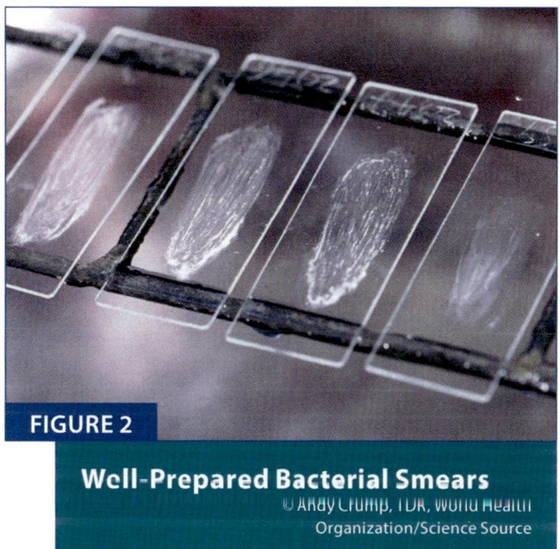

FIGURE 2

Well-Prepared Bacterial Smears
© Andy Crump, TDR, World Health Organization/Science Source

PRINCIPLES OF STAINING MICROORGANISMS

Because bacteria are so small and have such low contrast, bacterial smears are generally useless unless stained. The specific staining procedure that ought to be used is determined by the type of organism present in the smear and the types of questions you want to answer about the objective.

"Stains" are actually dyes. The ones that are commonly used in microbiology are usually based on a benzene ring and are composed of three basic parts:

- **Benzene ring** – Colorless but very versatile scaffold to which you may attach functional groups.

- **Chromophore** – The portion that imparts the color.

- **Auxochrome** – The charged portion that determines the binding properties. If this is omitted from the compound, you will have a chromogen (colored compound) and not a stain.

Methylene blue → (Ionization) → Cationic chromogen + [Cl]⁻

Picric acid → (Ionization) → Anionic chromogen + [H]⁺

FIGURE 1

Bacterial Staining. Examples of cationic (basic) and anionic (acidic) stains.

In this lab we use two major types of stains:

- **Acidic** – Stains that carry a negative charge (anionic). This includes compounds like eosin, picric acid, acid fuchsin, India ink, and nigrosin. These stains have the same charge potential as the cell walls and membranes of microorganisms. This means that the stain is generally not taken into the cell. Stains like eosin stay outside the cell where metabolites like acids can interact with them. Others, like India Ink, are used for negative staining techniques like capsule staining.

- **Basic** – These are positively-charged (cationic) stains that have a strong affinity for negatively-charged cellular components like bacterial cell walls and nucleic acids. Stains like crystal (gentian) violet and methylene blue are basic stains that will easily and quickly stain bacterial cell walls for visualization. Others, like safranin, are used as counterstains in differential staining procedures like the Gram stain. Basic fuchsin is used in acid-fast staining.

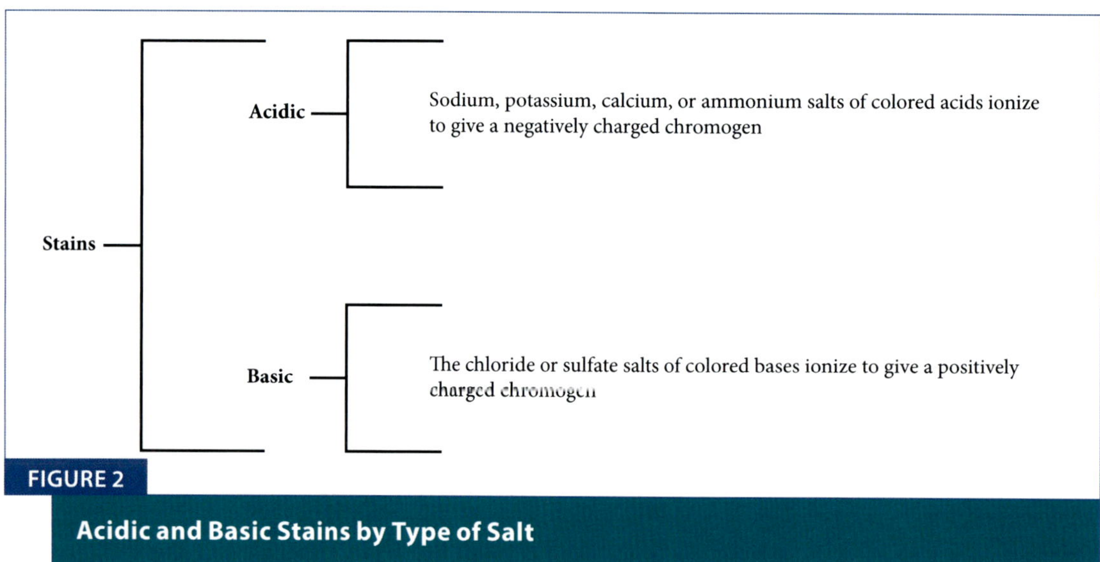

FIGURE 2

Acidic and Basic Stains by Type of Salt

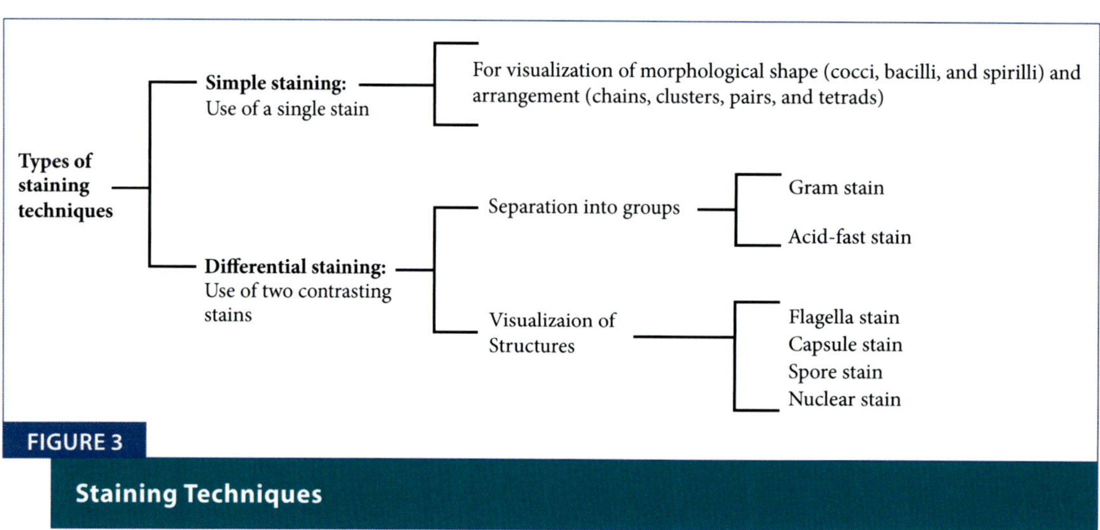

FIGURE 3

Staining Techniques

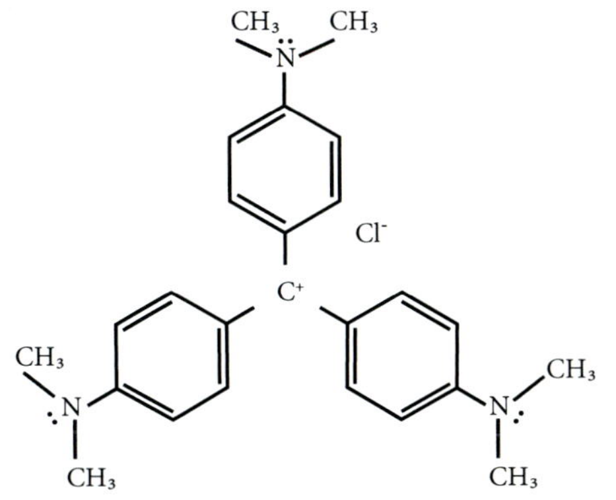

FIGURE 4

Crystal Violet

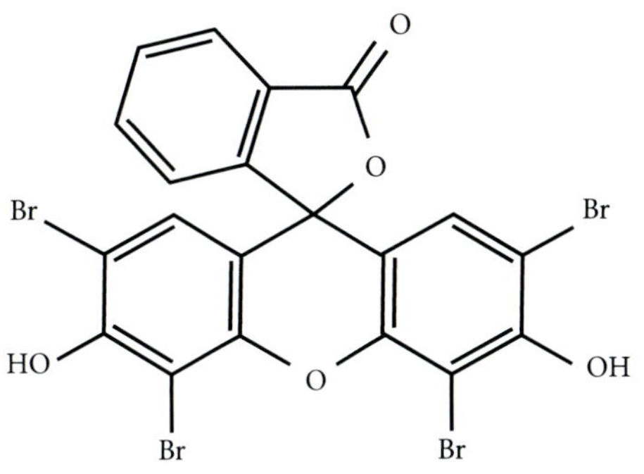

FIGURE 5

Eosin Y

These stains are used in many types of staining procedures, but the only two that we will use in this lab are **simple staining** and **differential staining.** Simple staining typically includes the use of one stain and is useful for observing cell **morphology** and **arrangement.** Differential staining is used to allow you to differentiate different types of bacteria, so they can be classified into different groups based on their biochemical anatomy.

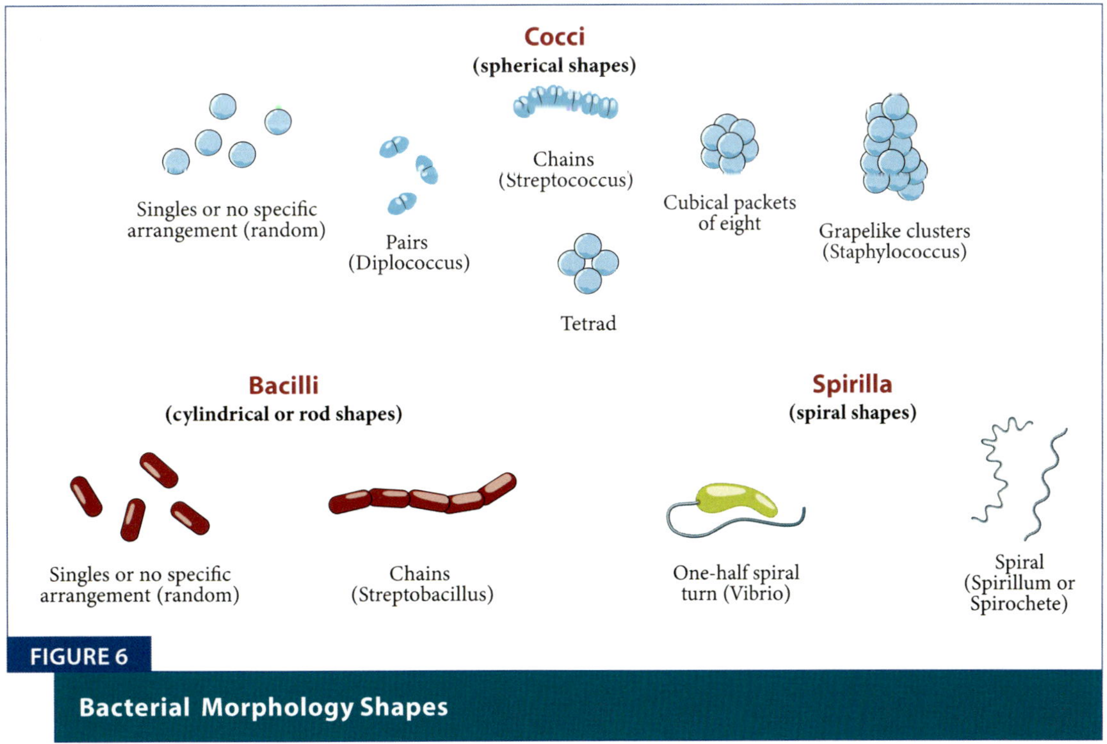

FIGURE 6

Bacterial Morphology Shapes

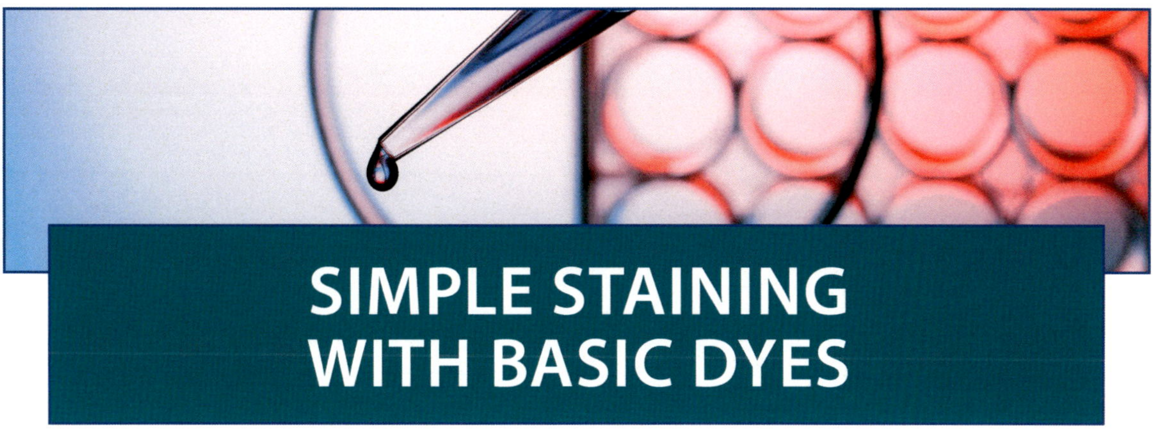

SIMPLE STAINING WITH BASIC DYES

Stains increase the contrast of microbial cells which allows us to more accurately and easily observe their gross properties. For the purpose of counting or identifying the morphology and arrangement of a particular bacterium, a simple staining procedure is adequate.

PROCEDURE

1. Obtain your prepared smear(s).

2. Place your smear on the staining rack smear side up.

3. Flood the surface of the smear with crystal violet or methylene blue and let it stand for 30 seconds to a minute (note how much time you give it).

4. Carefully rinse the excess stain off into the staining tray using a water rinse bottle.

5. Allow the slide to air dry (or help it along with fire).

6. Place on the microscope stage and add a drop of emersion oil directly to the smear.

7. Use the 100x objective and move the stage up until oil flows onto the objective lens.

8. Turn your light all the way up, your iris/diaphragm 60% closed, and look through the eyepieces while you make quick, clockwise, quarter turns of the fine focus knob.

9. Be careful that you don't overshoot the focal plane of your smear. If you go too far, you can cause the objective to come into contact with the slide. The objective lens is spring-loaded, so it can take a little abuse, but if you press too far, you will crack your slide and may scratch the lens coating which will ruin it.

10. Observe and report in your lab notebook the morphologies and arrangements of all the bacterial cultures being used in the lab.

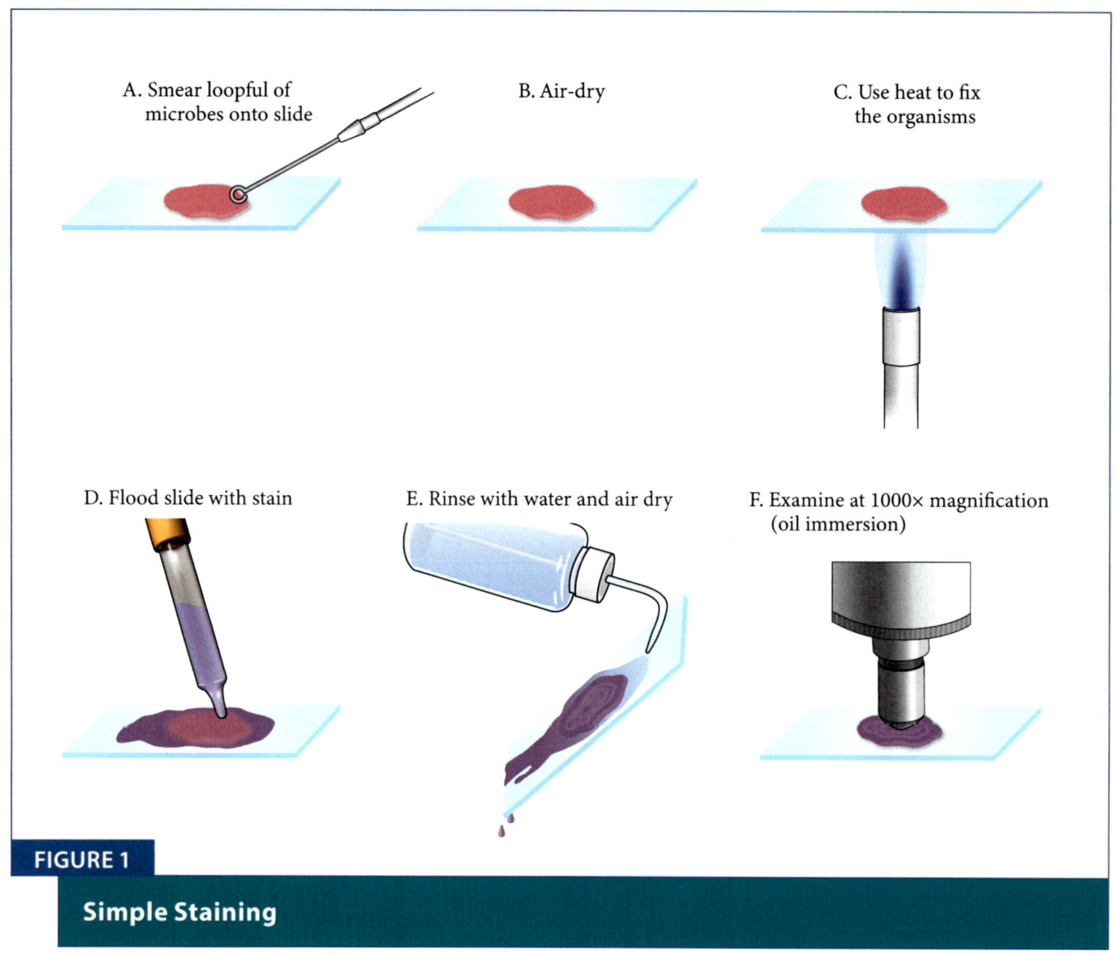

FIGURE 1

Simple Staining

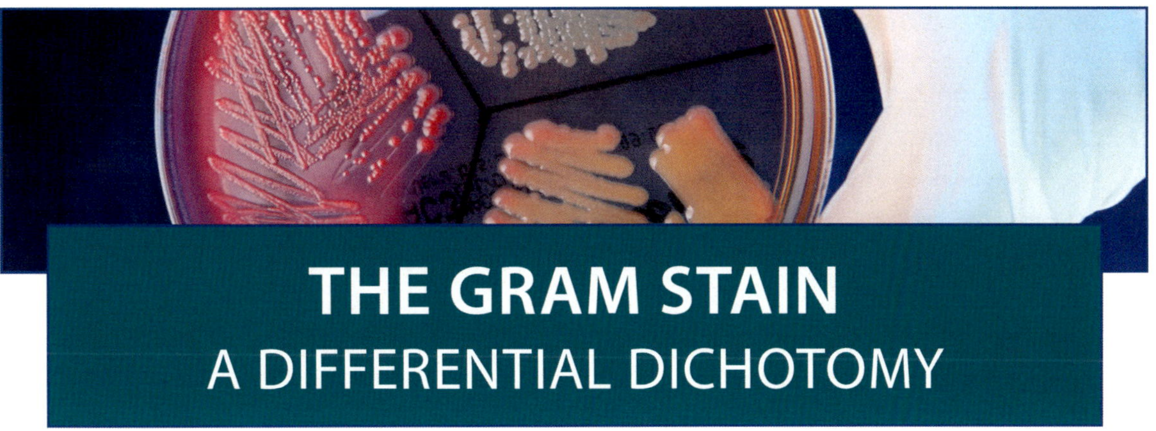

THE GRAM STAIN
A DIFFERENTIAL DICHOTOMY

The identification of microorganisms has been a major undertaking since microorganisms were first shown to exist. Microbes can be split into categories based on a number of physiological, biochemical, or anatomical properties, and these categories can have important clinical significance. In the case of bacteria, currently, the final word in **taxonomic identification** is DNA sequences, but unless you are willing to pay well above the going rate, it takes no less than two days to generate the data. In clinical applications where time, cost, and throughput can be critically important factors, physiological and morphological differences are the big players.

One of the most important steps in identifying a clinically-significant bacterium is to determine if it is **gram negative** or **gram positive.** The quickest and most cost-effective way to determine this is via the **Gram stain.** The Gram stain is a **differential staining** technique that can tell us about the arrangement of bacterial cell walls. When properly executed, the Gram stain will yield gram-positive (gram+) cells stained purple and gram-negative (gram–) cells stained red (sometimes can appear pink/orange if the microscope light isn't perfectly white). If performed on acid-fast bacteria, the Gram stain will cause the cells to come out both ways, which is called **variable staining.**

This procedure is named for its inventor, Dr. Hans Christian Gram, though it has undergone revision and refining since he published the first protocol. The basics have remained the same, however:

- **Primary stain** – A basic stain that has affinity for the **peptidoglycan** of bacterial cell walls. We will use crystal violet.

- **Mordant** – A compound that will combine with the primary stain to form a complex that is water insoluble. We will be using Gram's iodine.

- **Decolorizer** – A solvent that will wash away any primary stain and stain/mordant complex that is not bound up in peptidoglycan. We will be using 95% ethanol (EtOH).

⊙ **Counterstain** – Another basic stain with a coloration distinct from the primary stain. This will re-stain any cells that were made colorless by the decolorizer, but the color will be different from the primary stain. We will be using safranin.

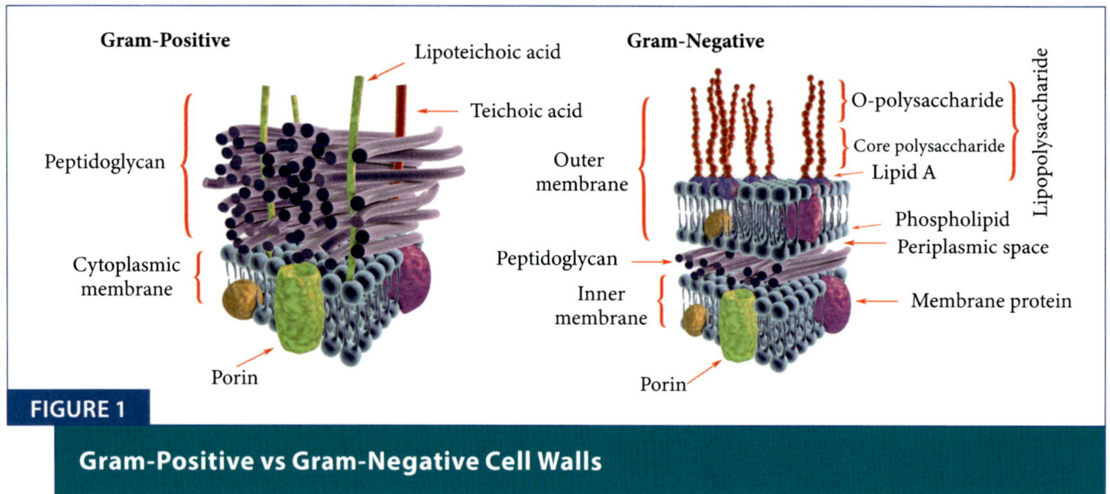

FIGURE 1

Gram-Positive vs Gram-Negative Cell Walls

The staining differences between gram+ and gram– are primarily a function of the thickness of their cell wall and whether they have an outer membrane. Gram+ cells have much thicker cell walls which consist of many overlapping layers of peptidoglycan cross-linkages which trap a significant amount of primary stain/mordant complex leaving them stained purple after decolorizing. Gram– organisms have a thinner cell wall, and it is buried below an outer membrane that acts as a gatekeeper that excludes primary stain. This means that virtually all the primary stain/mordant complex will be rinsed away by the decolorizor, leaving the cells colorless and ready to accept the counterstain.

PROCEDURE

You can proceed through these steps without waiting for the slide to cool or dry.

1. Create a bacterial smear. Be sure to burn off the packing oil and firmly heat-fix the smear because the ethyl alcohol (EtOH) will weaken the bond of the smear to the glass.
2. Place smear on the staining rack and flood the smear with crystal violet; let stand for 1 minute.
3. Rinse excess stain off with a rinse water bottle.
4. Flood with Gram's iodine and let stand for 1 minute.
5. Rinse mordant off with water.

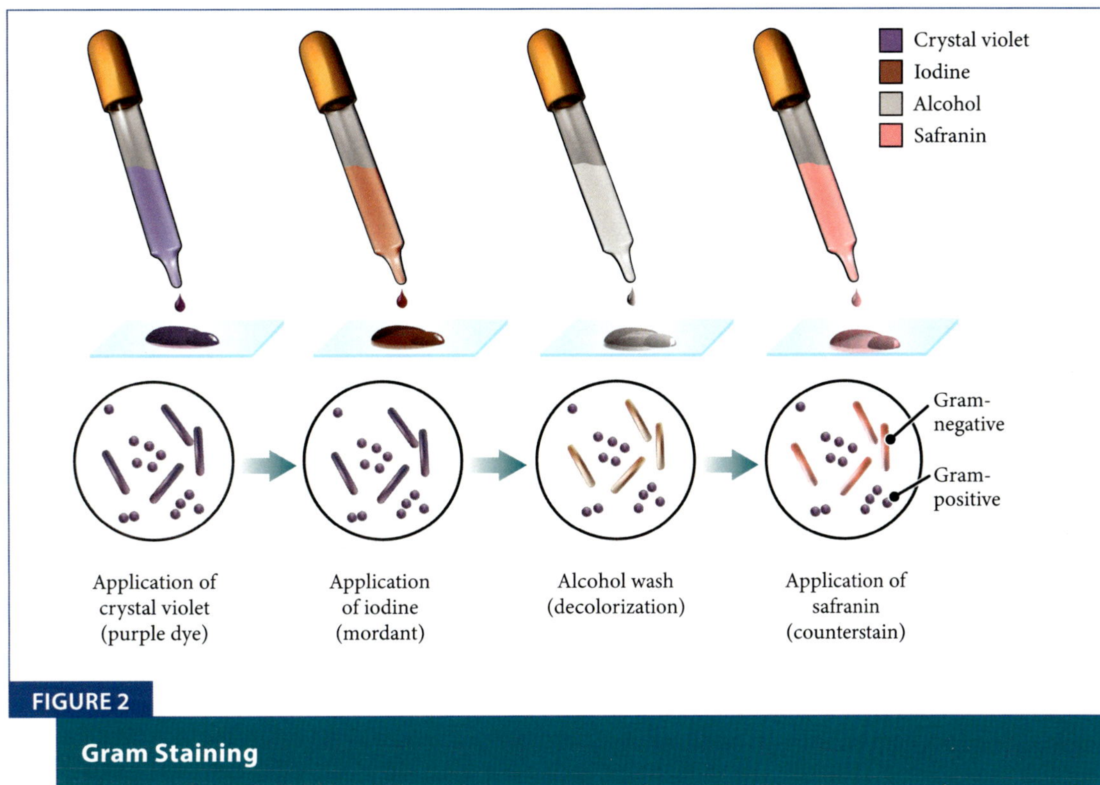

FIGURE 2

Gram Staining

6. Decolorize the smear by introducing EtOH drop-by-drop until the drops that fall off the slide are 95% clear of purple.

7. *Gently* rinse the EtOH off the slide using water.

8. Flood with counterstain and let stand for 1 minute.

9. Rinse the counterstain off with water.

10. Dry the slide and observe under 1000× oil immersion.

Finally, beware of performing the Gram stain on cultures that have been exposed to antibiotics such as penicillin or are older than 24 hours. These sorts of conditions can cause bacterial cells to alter the construction of their cell walls causing gram+ to appear gram– or gram-variable.

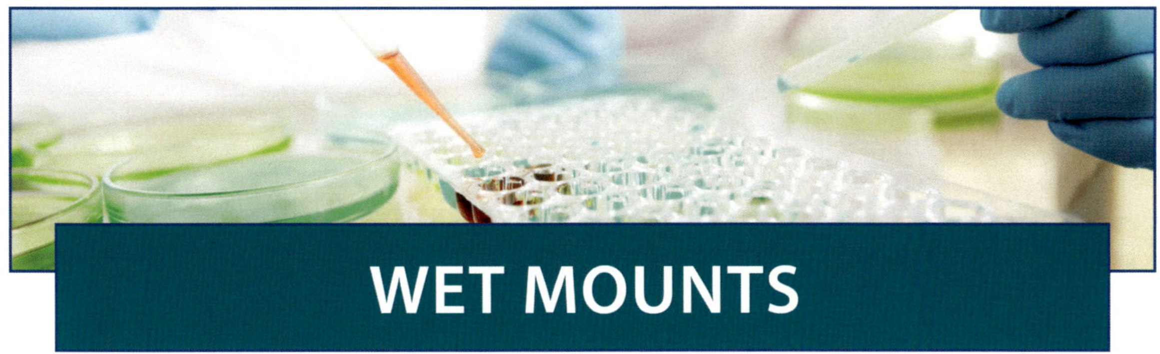

WET MOUNTS

Creating and staining bacterial smears causes the organisms to become affixed and stationary. They also become dead. Indeed, that's the main point of the procedure because we want them to stay put! There are, however, important aspects of microbes that can only be observed when the organisms are alive and free to move. To observe these properties, we will create **wet mounts.**

Wet mounts are exactly what they sound like: a liquid suspension of microbes on a slide. Bacterial cells are small, and their refractive index is close to that of water. These facts combine to make it challenging to observe them when wet mounted. Rise to that challenge, and you'll be rewarded with observations of **motility** (if the organisms have **flagella**), **binary fission, Brownian Motion** (the vibration of particles and cells as they are struck by water molecules), and the natural size and shape of the cells absent heat and chemical treatments.

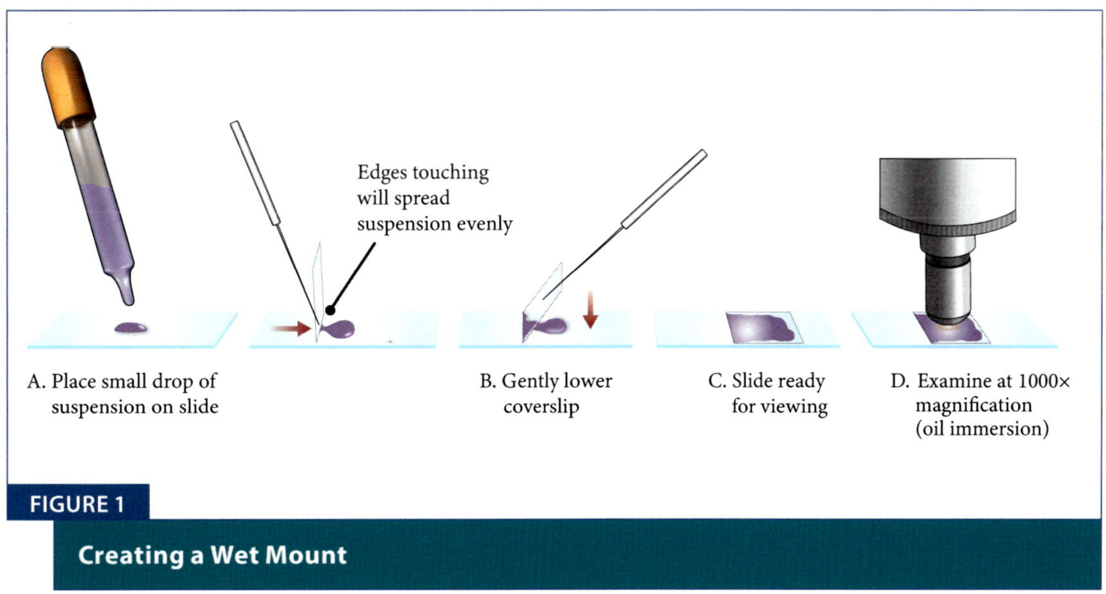

FIGURE 1

Creating a Wet Mount

PROCEDURE

1. Obtain a slide and burn off the packing oil.

2. Add 20–25 µl of a liquid bacterial culture to the slide.

3. Place a coverslip over the liquid.

4. Observe at 1000× under oil immersion.

SOME FINAL CONSIDERATIONS

Do one wet mount at a time because they dry out if you let them sit too long. Also, you'll be observing your wet mount using a standard brightfield microscope, but we would find it easier, though significantly more expensive, to make these observations if we were able to use phase-contrast microscopy.

SERIAL DILUTION PLATE COUNTS

It is often important to determine a quantitative estimate of how many microbes are present in a given environment. This is an especially important measurement to make during the manufacture of things like cosmetics and food or during key processes like wastewater treatment.

There are a lot of ways to estimate microbial abundance, and they all have their own strengths and weaknesses. We'll discuss just 6 different methods here to give a little context.

- **Direct microscopic counts** – Making direct counts of bacterial cells requires either a specialized slide called a **Petroff–Hausser counting chamber (hemocytometer)** (Figure 1) or an eyepiece lens that includes a grid. The PH counting chamber is a thick glass slide designed to suspend a known amount of liquid sample over a counting grid. An eyepiece with a grid can be used to enumerate liquid, fixed, or filtered samples, but the operator must calibrate the grid area at different magnifications using a stage micrometer. These methods of enumeration have the advantage of being quick, but they are not inherently sensitive to whether the cells are alive or dead.

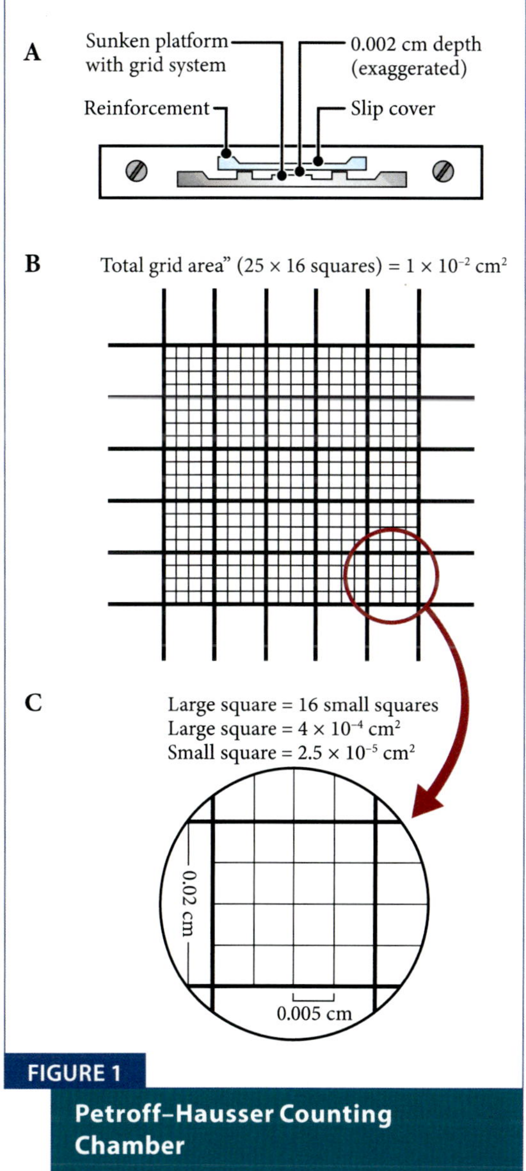

FIGURE 1

Petroff–Hausser Counting Chamber

- ⊙ **Coulter counter** – This is a bit of specialized equipment that can infer the number of cells in an electrolyte. It does this by passing fluid containing cells (or any particle) through a small aperture through which is flowing an electrical current (Figure 2). Cells are non-conductors so as they pass through the charged aperture, they increase the electrical resistance which is being recorded. This method has the benefit of being rapid and of not requiring the investigator to spend time at a scope making counts. Major disadvantages are that the machine can't distinguish live cells from dead cells and neither can it definitively tell you what is a cell and what is non-living particulate matter.

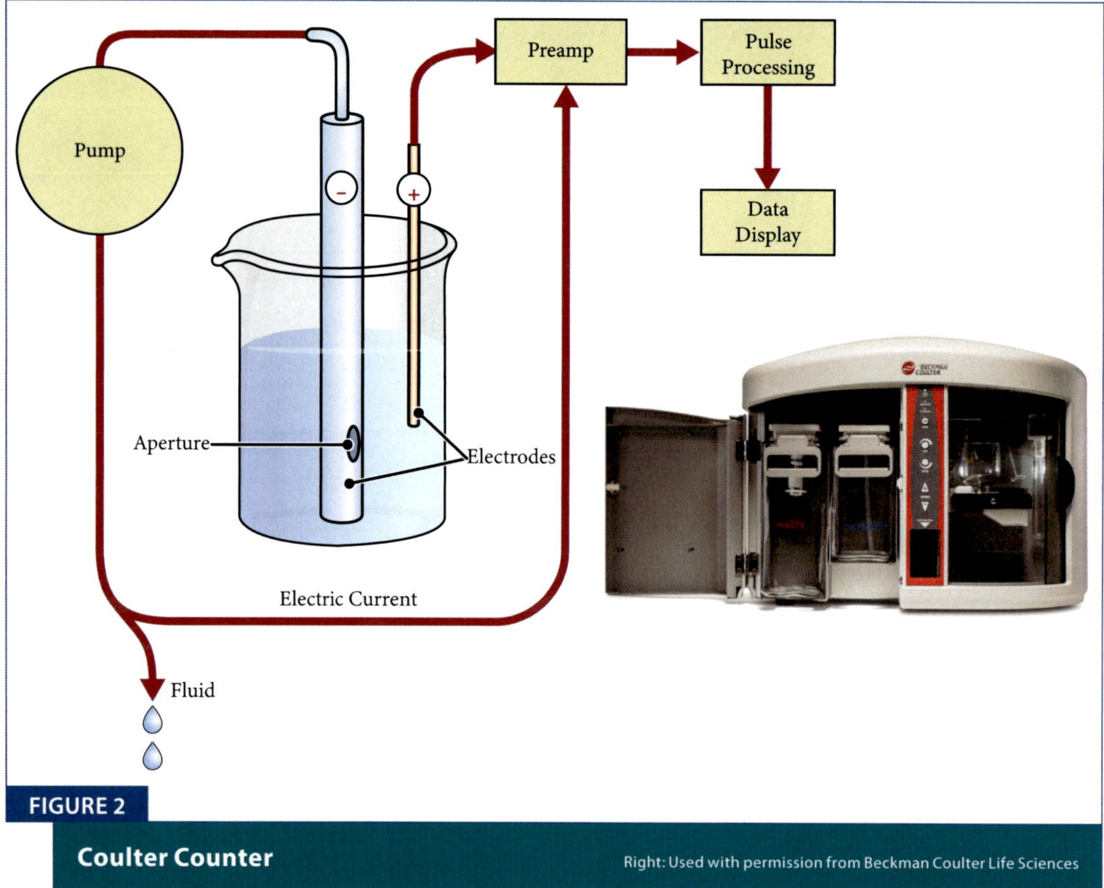

FIGURE 2

Coulter Counter   Right: Used with permission from Beckman Coulter Life Sciences

- ⊙ **Flow cytometer** – This is another piece of specialized equipment that detects cells as they pass through a narrow channel (Figure 3). A flow cytometer has a principle similar to that of a Coulter counter except that it uses lasers to detect the particles flowing through the detection channel. Flow cytometry has the benefit of using optical technology which can detect the size of each particle and can also be used to detect stains or natural pigments, and this can help cut out noise from non-cellular particulate matter. Advanced versions of these machines can also sort the cells based on their optical characteristics.

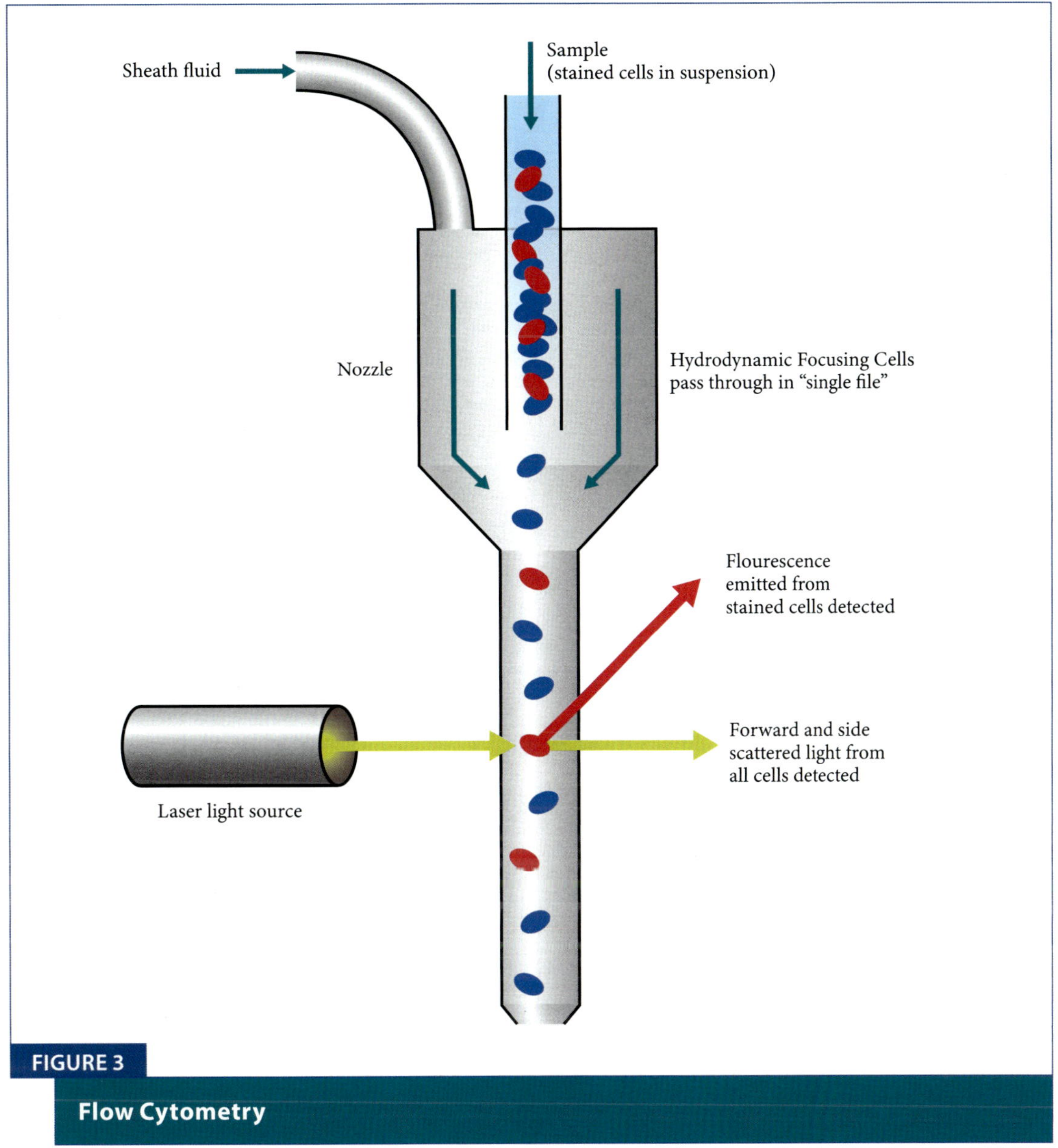

FIGURE 3

Flow Cytometry

⊙ **Spectrophotometric analysis** – This is another optical method wherein the **turbidity** (or **optical density**) of a microbial culture is used as an analog for cell number (Figure 4). Cells absorb and deflect light so that, as their population grows, the turbidity (cloudiness) of the culture increases, and this can be quantified by a spectrophotometer. These measurements are very quick, non-invasive to the sample, and inexpensive. The major disadvantages are that it is a relatively insensitive technique which means you need suspensions of 10 million cells or more to get detectable readings, and you aren't directly counting cells so you have to infer cell number. To be fair, however, turbidity can be linked to cell number if direct counts are performed concurrently with the optical measurements.

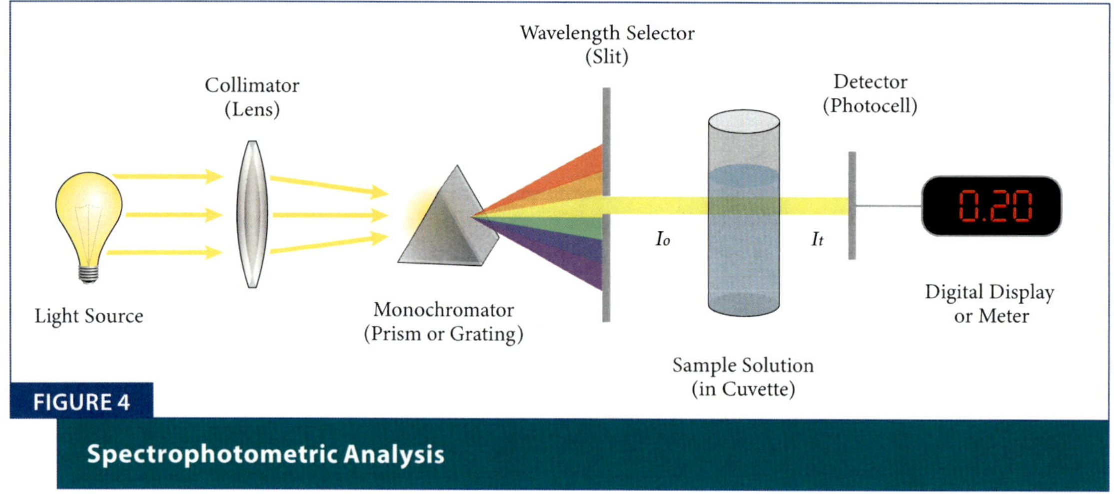

FIGURE 4

Spectrophotometric Analysis

⊙ **Chemical methods** – These shouldn't be considered to be quantitative measurements since they aren't always directly linked to the number of cells present. These methods include measurements of protein, fat, and DNA concentrations (spectrophotometry again) or dry weight, oxygen consumption (special electrodes or wet chemistry) (Figure 5), or gas (like H_2) production (gas chromatograph).

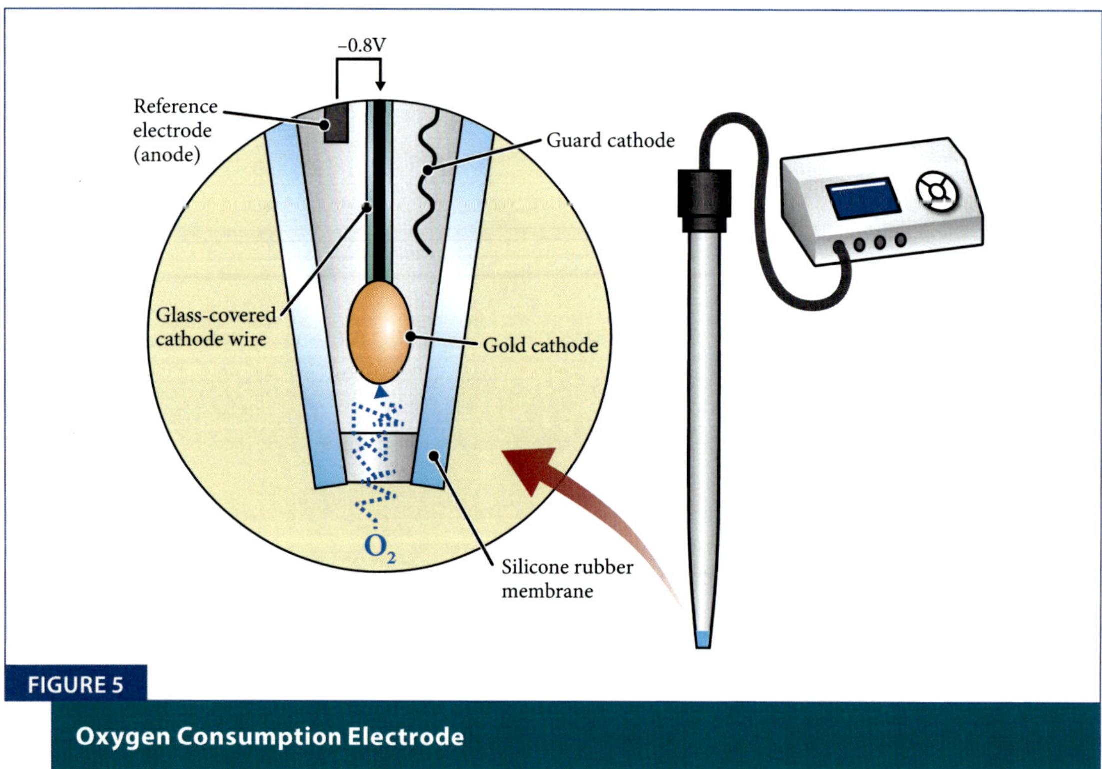

FIGURE 5

Oxygen Consumption Electrode

⊙ **Serial dilution plate counts** – This is the method that you will be employing this week (Figure 6). This is the only method listed here which can provide an estimate of the concentration of **viable cells**. In many medical and sanitary applications, it is the microbes that are metabolically active and capable of reproducing (viable) that matter. It's time consuming and requires more disposable materials and media than the other methods, but it can also yield discrete colonies that can be used to create pure cultures.

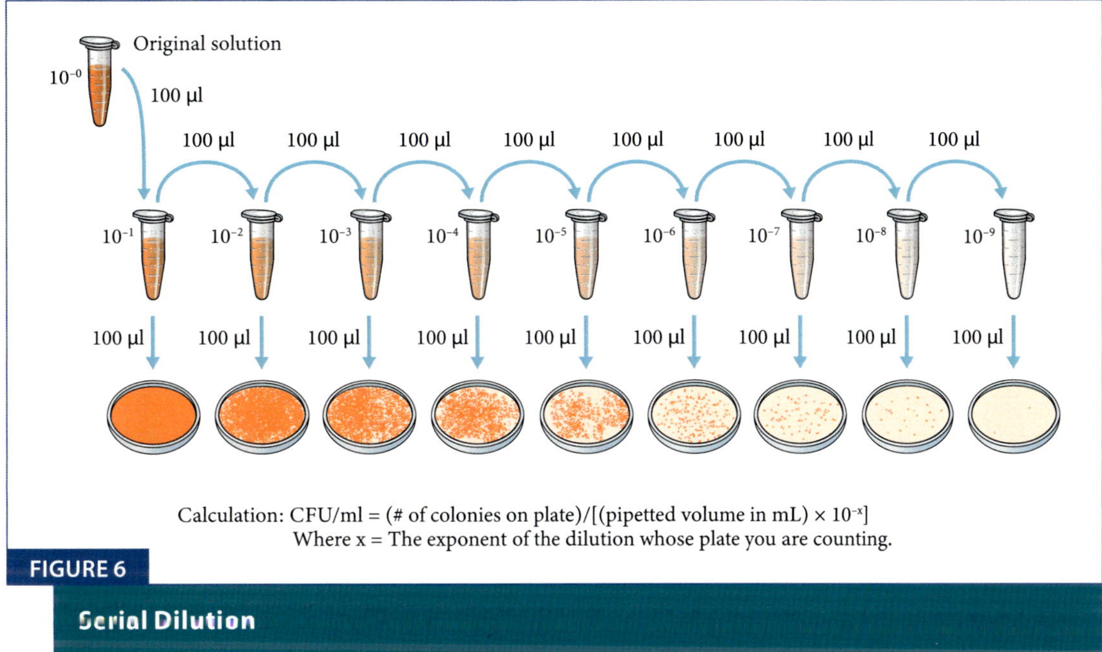

Calculation: CFU/ml = (# of colonies on plate)/[(pipetted volume in mL) × 10^{-x}]
Where x = The exponent of the dilution whose plate you are counting.

FIGURE 6

Serial Dilution

PROCEDURE

1. Obtain a tube of liquid bacterial culture, enough tubes of sterile saline to do all the dilution steps, a pipette, some pipette tips, and a stack of TSA plates sufficient to plate the required dilutions.

2. Note the dilution factor of your given bacterial culture. Label it 10^0 if it is undiluted. Label the rest of the tubes in descending exponents from what you labeled the first tube, i.e. 10^{-1}, 10^{-2}, 10^{-3}, **ALL THE WAY THROUGH TO 10^{-9}.**

3. Vortex the bacterial culture to re-suspend any cells that settled out.

4. Open the bacterial culture, extract 100 µl, and transfer it to the first (e.g. 10^{-1}) sterile saline tube.

5. Close both tubes and vortex the 10^{-1} tube.

6. Re-open the 10^{-1} tube and transfer 100 µl to the 10^{-2} tube.

7. Close the tubes, vortex the 10^{-2} tube, and proceed through the rest of the dilutions.

8. Once all dilutions have been completed, **OPEN EACH TUBE IN TURN** and transfer 100 µl into a single TSA plate, labelling the plates with the dilution factors as you go.

9. Discard all dilutions in biohazard containers and incubate the plates at 37° C for 24 hours.

10. After incubation, choose *a single plate* to count, being sure the plate has a number of colony forming units (CFU) within 30–300.

11. Convert your CFU count to CFU/mL in the original culture according to the following equation:

CFU/ml = (# of colonies on plate)/[(pipetted volume in mL) $\times$ 10^{-x}]

Where x = The exponent of the dilution whose plate you are counting.

For example, you do your dilution according to the protocol given in this exercise. Do plating and incubate as described. After incubation, you find that that 10^{-4} plate has an appropriate number of colonies in it for counting. You count and find 222 CFU. The math would be as follows:

CFU/ml = 222/(0.0001 $\times$ 0.100) = 22,200,000

MICROBIOLOGICAL ANALYSIS OF FOOD & WATER

Humans have been using microbes to preserve and modify foods for as long as humans have been human. Archaeologists have determined that beer was being brewed in ancient China, as many as 5000 years ago (~3000 BC). Microorganisms are a necessary part of many foods today as well. It is yeasts and bacteria that turn grain into beer, milk into yogurt, and smoked ground sausage into pepperoni or salami. Even so, it is equally true that potentially harmful microbes will also grow in/on foods meant for human consumption. A common contaminant in foods that may indicate greater risk are **coliform bacteria,** which are small gram-negative bacilli capable of fermenting lactose. Coliforms are considered an indicator of the possible presence of other dangerous bacteria because they (coliforms) are closely associated with fecal contamination and they are easy to detect. The colons of warm-blooded animals have extremely high concentrations of coliforms (and many other bacteria), so fecal contamination of food or water can be indicated by the presence of coliforms.

Something similar is true with regard to natural waters. There is a wide array of microorganisms that inhabit natural waters and perform essential ecosystem functions. There are **eukaryotic** microbial algae that capture energy from the sun, use it to fix carbon from the atmosphere into complex carbohydrates and proteins, which then serve as food for higher **trophic levels.** There are nitrifying bacteria that convert nitrate (NO_3^-) into nitrite (NO_2^-), ammonia (NH_3), and nitrogen gas (N_2). Some of the same contaminants that you may find in food also act as indicators of fecal contamination in water bodies. When waters have been contaminated by fecal matter, there is a chance that there may be more dangerous species present, such as *Vibrio cholerae,* which is the cause of cholera. Coliforms are an acceptable metric to use in temperate climates where soil temperatures are typically much lower than mammalian body temperatures, but coliform counts will not help you much in warm and wet tropical regions. Here, species like *E. coli* may be found in the soil and water of areas with no provable fecal contamination.

In this laboratory exercise, you will be using one of the most common methods for quantifying viable microorganisms that are associated with food products and natural waters: serial dilution agar plating. You learned this skill in a previous lab exercise, and now you will put your skills to the test using prepared foods purchased from a local deli counter and samples from natural waters in and around Columbia, SC.

PROCEDURE

1. Obtain six tubes of sterile saline, a food sample, a water sample, six TSA plates, and four each of Eosin Methylene Blue (EMB) and MacConkey (MCA) agar (see page 78 for description).

2. Label your saline tubes and plates with your name, section number, your food sample, your water sample, and the dilution factors 10^0–10^{-2}.

3. **Come to an agreement with your laboratory colleagues on the standardized sample size for the various foods.** You may use the available balances to measure out your sample.

4. Place the analytical fraction of your food sample into the 10^0 saline tube and vortex it to homogenize it as much as possible.

5. Allow large particles to settle for a few moments and immediately transfer 1 mL of the liquid from the 10^0 tube and place it into the 10^{-1} tube. Vortex the tube 10^{-1} and repeat the procedure again into the 10^{-2} tube.

6. Once your serial dilutions are done, vortex them one more time to keep any cells suspended in the liquid, then immediately transfer 100 µL into the corresponding TSA and EMB plates and spread them using the glass bar.

7. Repeat this procedure again for your water sample, using MCA instead of EMB.

8. Incubate all plates at 37° C for 24 hours.

9. After incubation, observe the plates looking for visible colonies. If there is a plate that falls between 30–300 colonies, count those colonies and use the count to calculate the number of viable cells and coliforms (if present) in the analytical sample (**number of cells per g or mL**).

10. Collect results from the other food and water samples for your lab notebook.

CULTURE MAINTENANCE

A major focus of this laboratory course is to educate future healthcare and pharmaceutical professionals about how to cultivate desirable microorganisms and reduce or eliminate harmful ones. It is therefore important for you to know something about what microorganisms require to live.

The microbes you've been working with are living things, and that means that they need nutrients, energy, and water to continue living. Microbes are very diverse, so not all microorganisms have the same requirements. The parameters that some microorganisms require to thrive are bad enough for other organisms that they will die if exposed to them. It is therefore important that we know the general requirements for an average organism *and* the specific requirements for whatever is our specific organism.

In general, microorganisms require the following:

- ⊙ **Water** – Every living thing we know of requires some amount of water to live and reproduce. The cytoplasm within cells is approximately 80% water and 20% dissolved substances. Microorganisms also need water outside the cell to facilitate the movement of low molecular weight nutrients across the cell membrane.

- ⊙ **Carbon** – Considered a **macronutrient,** it is the element most commonly found in living things and is half of the definition of what constitutes **organic compounds.** The dry weight of living things, micro or macro, is predominantly carbon. It is important to note how organisms acquire their carbon:

  - ◎ **Autotroph** – *Auto* = "self, same, or spontaneous" and *trophic* = "pertaining to food." Organisms capable of autotrophy obtain their carbon from inorganic carbon in their environment (usually carbon dioxide) and **fix** it into organic compounds for use in their cells. These organisms are not human pathogens, except in the case of *Clostridium difficile,* which has some autotrophic genes and has been shown capable of growth with $CO_2 + H_2$ as a sole carbon source.

  - ◎ **Heterotroph** – *Hetero* = "different or other." Organisms capable of heterotrophy obtain their carbon via organic carbon (previously fixed from CO_2) in their environment. All human pathogens (with the exception of *Clostridium difficile*) fall within this class. Such pathogens obtain their carbon from *you*.

○ **Mixotroph** – *Mix* = "mingling". Mixotrophic organisms are capable of both autotrophy and heterotrophy and will engage either of them depending on environmental conditions. Many bacterial and some microalgal species are capable of mixtotrophy. Even some higher plants (macrophytes) such as venus flytraps and perhaps even animals such as the oriental hornets exhibit a level of mixtotrophy.

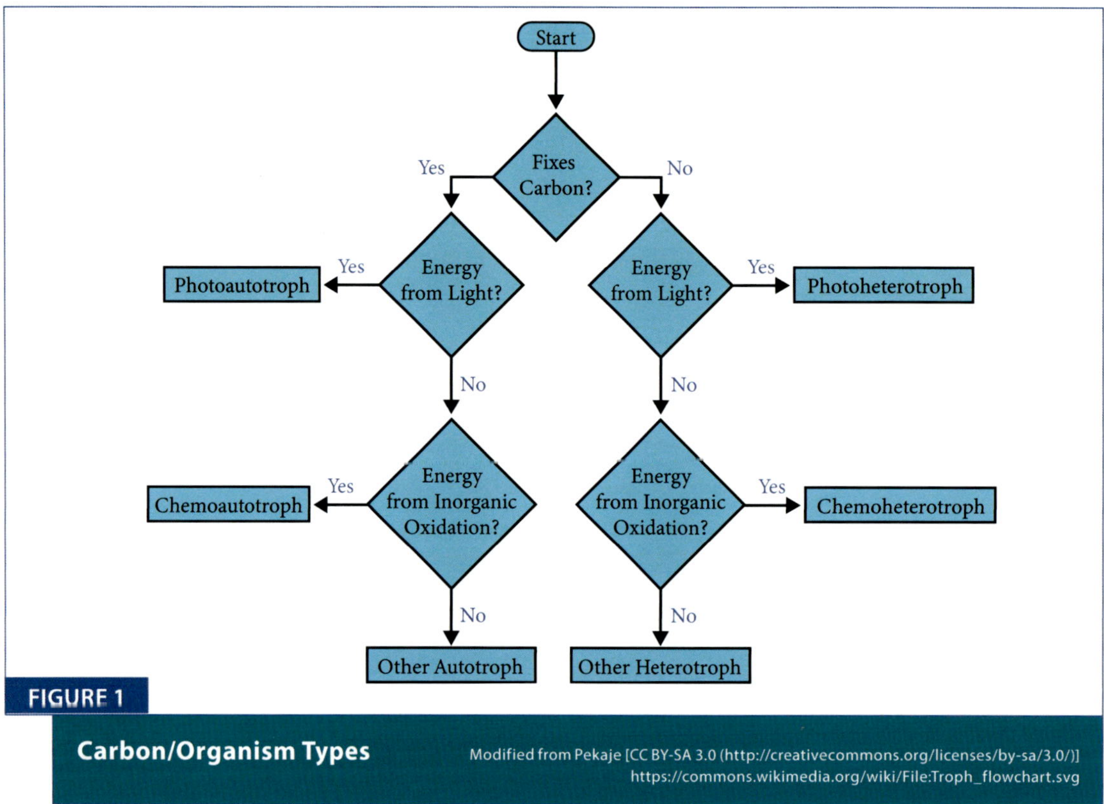

FIGURE 1

Carbon/Organism Types   Modified from Pekaje [CC BY-SA 3.0 (http://creativecommons.org/licenses/by-sa/3.0/)] https://commons.wikimedia.org/wiki/File:Troph_flowchart.svg

○ **Nitrogen** – This is another macronutrient that, in terms of mass, is perhaps the second most abundant element in microorgansms. Nitrogen is used in many essential biological compounds, especially amino acids (proteins) and nucleic acids. Proteins are the workhorses of the cell, performing essential duties, especially as enzymes that lower the **activation energy** of key chemical reactions. Nucleic acids are the medium of the genome wherein is stored all of the information necessary for the life of the given organism (DNA). They also comprise the templates for protein translation (mRNA) and are key components of the ribosomes that do the translation (rRNA), combining the amino acids into a peptide chain (tRNA). Nucleic acids can also be used like **adenosine triphosphate** (ATP) as an energy currency within the cell. There are **nitrogen fixers** capable of splitting the triple covalent bond in N_2 and making NH_3, **nitrifyers** who reduce oxidized nitrogen species, and **denitifyers** who turn bioavailable nitrogen back into N_2 gas. The nitrogen species you need to supply depends on the organism in culture: N_2 gas, NH_3, or NO_2^-, NO_3^- salts.

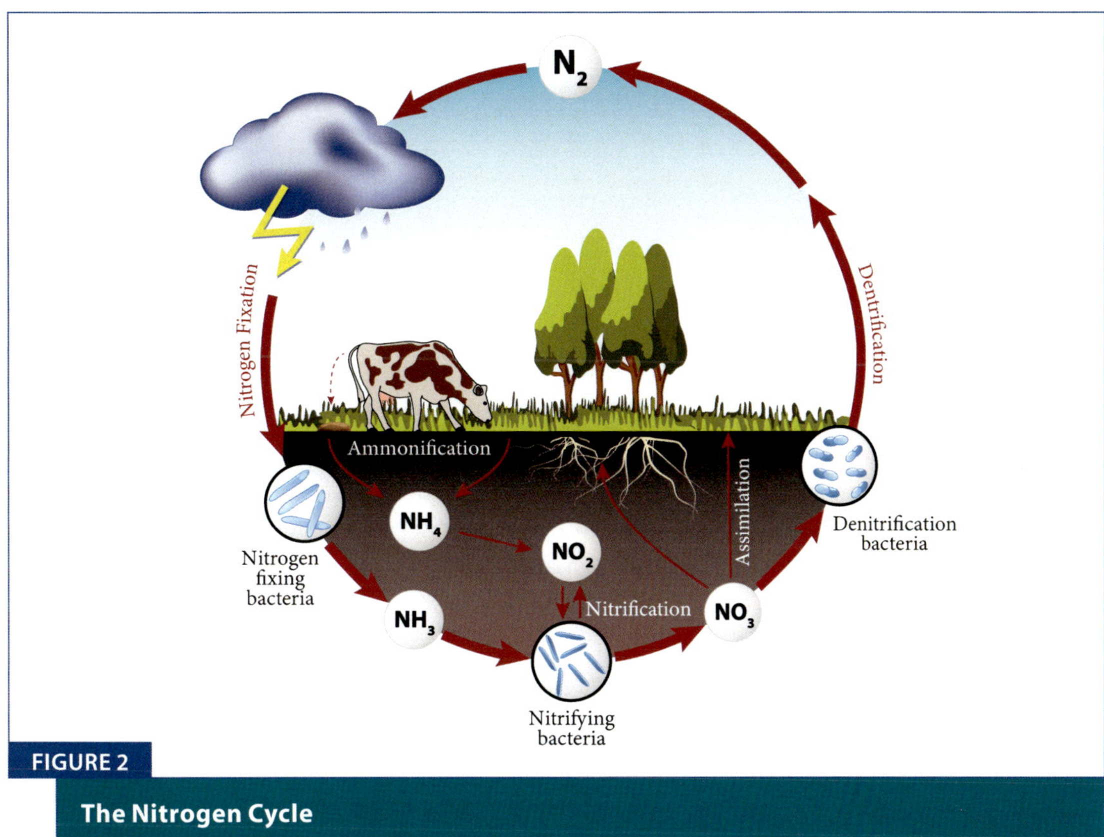

FIGURE 2

The Nitrogen Cycle

- **Non-metallic elements** – These are mainly P and S, which are also macronutrients. Phosphorus is a key player in the energy economy of the cell and is an essential part of genomic structure. S is a common constituent of proteins, a component of some lipids, and is important in nitrogen metabolism. These are supplied to microbes as P salts, amino acids, sulfate, or elemental sulfur.

- **Metals** – These include elements like Na^+, K^+, Mg^{2+}, Ca^{2+}, Mn^{2+}, Fe^{2+}. Fe^{3+}, Cu^{2+}, Zn^{2+}. Several of these function as essential cofactors in the operation of enzymes. Others function as osmoregulators or in the transport of electrons during metabolism. Most metals are considered **micronutrients** and are usually supplied via metal salts.

- **Vitamins** – These are organic substances, usually supplied in minute amounts, that function as metabolic coenzymes facilitating efficient metabolism. The larger the profile of vitamins an organism requires be supplied in the medium, the less flexible and more **fastidious** it is. Many human pathogens are fairly fastidious.

- **Energy** – Most activities in the cell require energy input to proceed. Active membrane transport, amino acid synthesis, DNA transcription, protein translation, and motility all require a large amount of energy. Organisms can be classed according to how they obtain their energy:

◎ **Phototrophs** – These obtain their energy from solar radiation. These organisms use systems of pigments that capture energy from certain photons and use that energy to excite electrons from which the energy can be harvested. These organisms will therefore require light in order to grow.

◎ **Chemotrophs** – These organisms obtain their energy by oxidizing chemical compounds, either organic or inorganic. The most common organic energy source is dextrose (glucose), but there are many microorganisms that can use inorganic compounds such as H_2S or $NaNO_2$.

⊙ **Gaseous atmosphere** – It is common for an organism to require atmospheric oxygen (**obligate aerobe, microaerophile**), but it is also common for organisms to thrive without it (**obligate/facultative anaerobes, aerotolerant**). Other organisms, such as many human pathogens, require some specific fraction of their gaseous environment to be CO_2 up to perhaps 3%–5%. For some organisms, it is a matter of metabolic terminal electron acceptors and for others, like photoautotrophic microalgae, it's a matter of obtaining carbon for fixation. For obligate anaerobic organisms, it is a matter of excluding the atmospheric oxygen that leads to the formation of deadly peroxide and superoxide.

⊙ **Temperature** – Every organism has an **optimal temperature** at which its compliment of enzymes runs most efficiently and allows the organism to grow at its highest theoretical rate. Every organism also has a **temperature range** over which it can grow at some rate less than the optimum. Terms like "low" and "high" are relative with regard to temperature. There are microorganisms capable of thriving in the superheated seawater deep in oceananic volcanic vents, and there are microorganisms that thrive between layers of Antarctic ice. What is a high temperature for one organism is low for another and vice versa. Low temperatures slow enzymatic reactions, slowing metabolism and growth. High temperatures can **denature** enzymes, rendering them useless and killing the organism.

⊙ **pH** – Every microorganism has some **optimal pH** which will allow maximal growth rate and a **pH range** over which growth can occur but at a rate slower than the optimum. pH (not Ph or PH) is the negative log of the concentration of hydrogen ion ($-\log [H^+]$) in a solution. The pH of the environment can greatly impact the speed of enzymatic reactions and the generation of ATP, which proceeds via **chemiosmosis.** Like with temperature, discussion of "high" and "low" pH values is discussion of relative amounts. pH < 7 = acidic, pH = 7 = neutral, pH > 7 = alkaline (basic). A pH too far outside the range for a given organism can cause denaturation of enzymes and death.

CULTURE MAINTENANCE

The maintenance of microbial cultures means managing their environment so as to keep them living and reproducing. In practice, this generally means the creation of fresh growth media, subculturing organisms in/onto it at regular intervals, and incubating the subcultures at an appropriate temperature.

These subcultures are required because as microorganisms grow on/in their medium, they use up nutrients and they export metabolic wastes that will act as poisons if they reach too high a concentration. If you wait too long to subculture a microorganism, you risk having a population crash wherein the vast majority of the cells either die or enter a resting state from which it may or may not be trivial to wake them. "Too long" could be as little as 48 hours for some species and a matter of 6+ months for others. The microbiologist must be attuned to the appearance of their microbial cultures and act accordingly.

A key consideration during culture maintenance is that you maintain aseptic technique throughout. Every time you open your microbial culture, you risk contaminating it with one or more other organisms; therefore, you should keep your subculturing routine to the minimum required for your purposes. You must also always be practicing aseptic technique and resist the temptation to get complacent and let your brain switch to autopilot.

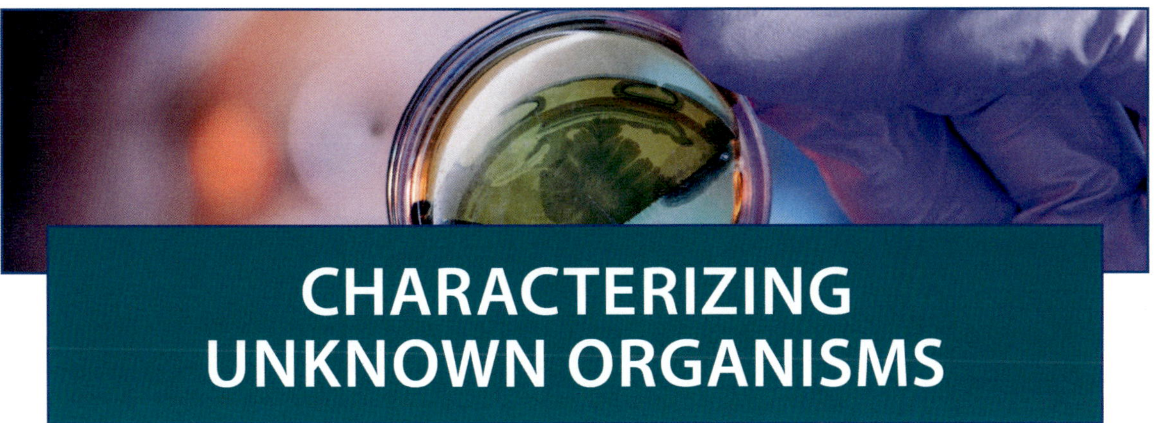

CHARACTERIZING UNKNOWN ORGANISMS

The characterization and identification of unknown microorganisms have been major foci within microbiology, medicine, sanitation, and food manufacture/preparation since visual evidence for the existence of microbes was first published in the mid-1600s AD. The delineation began in earnest during the early-mid 1800s AD as the old **theory of abiogenesis** was gradually supplanted by the **theory of biogenesis,** and many medical professionals began to focus on identifying the material causes of infectious disease.

Traditionally, much of the characterization and identification of microorganisms has been based on microscopic appearance, the manner in which they are stained by various dyes, or how they responded to or changed the appearance of various types of nutrient media. These techniques have the benefits of more than a century's use, rapidly generating data, and in most cases, being less expensive than more modern genomic techniques. These more "low tech" methods still enjoy considerable use in clinical applications where speed of data acquisition, high sample throughput, and per sample cost are essential.

The classification/delineation of organisms is called **taxonomy** which comes from the Greek *taxis* "arrangement" and *–onomy* "method." Humans have been doing taxonomy of plants and animals for as long as we have a history (and probably before). The current standard for bacterial taxonomy is *Bergey's Manual of Systematic Bacteriology, 2nd Edition,* which is a 5-volume (8-part) set of books that delineates bacteria into 33 groups rather than the traditional arrangement of phylum, class, order, and family. In the two most recent editions, the focus of the work has begun to shift away from "artificial" classification systems based on **phenotype** and toward what the authors call "natural" systems based on **genotype (phylogeny)**. Even so, it is acknowledged in the opening chapters of Vol. 1 that classifying bacteria genotypically is fraught with difficulty, especially as we try to classify the higher orders of the taxonomic hierarchy. It should therefore not be surprising when you learn that phenotypic taxonomy still plays a essential role in the taxonomy laid out in *Bergey's* and in microbiology in general.

In this laboratory course, we will be following in the footsteps of the brave men and women who have blazed the trail for us, and we will do so by employing phenotypic classification methods. These are the methods most commonly used in the practice of medicine, and they are the ones that you are most likely ever to come into direct contact with in whatever career you find yourself in.

We will use several aspects of microbial phenotype including colony morphology, cellular morphology, cell wall structure, physiology, and biochemistry to classify a set of bacterial organisms. By this point, you have developed sufficient knowledge of aseptic technique, sample preparation, staining, and specialized growth media that you are ready to move to the next challenge: *Identifying a bacterium whose identity you don't already know.*

A common way of approaching this type of problem is to evaluate the organism's compliment of **enzymes.** Enzymes are arguably the one class of molecule that allows life to exist. They are mostly a subset of the proteins, and they make it possible for many chemical reactions to occur and at temperatures at which life can exist. Different organisms may have different enzymes and may therefore be capable of different types of **cellular metabolism.** There are two major types of enzymes:

- **Intracellular/endoenzymes** – These function inside cells and are responsible for the synthesis of new cell components and the extraction of energy.

- **Extracellular/exoenzymes** – These are made inside the cell but function *outside* the cell. We will look specifically at **hydrolytic exoenzymes** whose function is to use water to break useful high molecular weight compounds into smaller bits that can be transported into the cell.

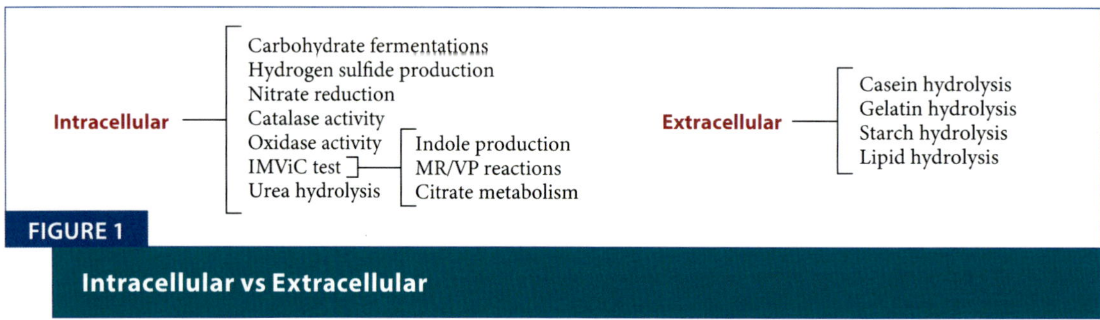

FIGURE 1

Intracellular vs Extracellular

You will find a list in Table 1 that contains every organism that your instructor will hand out. Beware, however, that not every organism in the table will be handed out this semester. Your job will be to, over the next few weeks, subject your given organism to a battery of tests and use the data that you generate to write a paper in which you identify your organism, recount how you came to the given ID, why you are sure it is *this* organism and not another one, and you'll present relevant details about the organism to the reader to provide a context for why your organism matters.

As you work, keep in mind that no single test can, by itself, give you a positive identification. You will need all of the tests, and you should therefore view your dataset as a profile of the organism. Your job will be to compare your organism's profile to the profiles in Table 1 and make a determination about what organism you have.

Biochemical Profiles

TABLE 1

ORGANISM	MORPHOLOGY	GRAM STAIN	LACTOSE	DEXTROSE	SUCROSE	H$_2$S PRODUCTION	INDOLE PRODUCTION	NO$_3$ REDUCTION	MR REACTION	VP REACTION	CITRATE USE	CATALASE ACTIVITY	OXIDASE ACTIVITY	UREASE HYDROLYSIS	CASEIN HYDROLYSIS	GELATIN HYDROLYSIS	STARCH HYDROLYSIS	LIPID HYDROLYSIS	HEMOLYSIS
Enterococcus faecalis	Coccus	+	A	A	A	-	-	-	+	+	-	-	-	-	+	-	-	-	γ
Lactococcus lactis	Coccus	+	A	A	A	-	-	-	+	-	-	-	-	-	+	-	-	-	γ
Micrococcus luteus	Coccus	+	-	-	-	-	-	±	-	-	-	+	-	+	-	S+	-	-	α
Staphylococcus aureus	Coccus	+	A	A	A	-	-	+	+	±	-	+	-	-	-	R+	-	+	α
Alcaligenes faecalis	Coccobacillus	-	-	-	-	-	-	-	-	±	+	+	+	-	-	-	-	-	α
Bacillus cereus	Bacillus	+	-	A	A	-	-	+	-	±	-	+	-	-	+	R+	+	±	β
Bacillus thuringiensis	Bacillus	+	-	A	-	-	-	+	+	±	-	+	+	+	+	R+	+	-	β
Corynebacterium xerosis	Bacillus	+	-	A±	A±	-	-	+	-	-	-	+	-	-	-	-	-	-	β
Escherichia coli	Bacillus	-	AG	AG	A±	-	+	+	+	-	-	+	-	-	-	-	-	-	α
Klebsiella aerogenes	Bacillus	-	AG	AG	AG±	-	-	+	-	+	+	+	-	-	-	-	-	-	α
Klebsiella pneumoniae	Bacillus	-	AG	AG	AG	-	-	+	±	±	+	+	-	+	-	-	-	-	α
Proteus vulgaris	Bacillus	-	-	AG	AG±	+	+	+	+	-	±	+	-	+	-	R+	-	-	α
Pseudomonas aeruginosa	Bacillus	-	-	-	-	-	-	+	-	-	+	+	+	-	+	R+	-	+	β
Pseudomonas fluorescens	Bacillus	-	-	-	-	-	-	+	-	-	+	+	-	-	-	-	-	-	γ
Salmonella typhimurium	Bacillus	-	-	AG±	A±	+	-	+	+	-	+	+	-	-	-	-	-	-	α
Shigella dysenteriae	Bacillus	-	-	A	A±	-	±	+	+	-	-	+	-	-	-	-	-	-	β

Note: ± = variable reaction, AG = acid and gas, S = slow, R = rapid

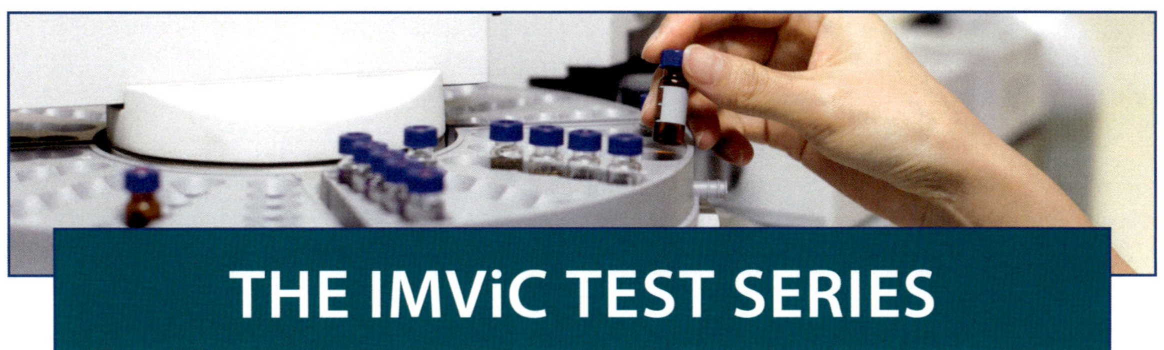

THE IMViC TEST SERIES

The **I**dole, **M**ethyl red, **V**oges–Proskauer, **in** Citrate series is a group of tests that use specialized growth media for the purpose of further classifying coliform bacteria. While this testing series is particularly good at delineating members of the **Enterobacteriaceae,** it can also be useful outside that context as you will see by the end of this laboratory exercise.

The Enterobacteriaceae are a family of bacteria that share the traits of being **mesophilic,** short, gram– bacilli that do not form endospores. The IMViC testing series was initially designed to assist clinical professionals in classifying samples containing Enterobacteriaceae so they could determine which they were:

- ⊙ **Normal intestinal flora** – This category would include organisms in the genera *Escherichia* and *Enterobacter,* for instance, which are **saprophytes** within the latter portion of your intestinal tract. While there may be pathogenic strains within some of the species, most are either **commensal** or **mutualistic** with their human hosts.

- ⊙ **Occasional pathogens** – This could include species within the genera of *Klebsiella* or *Proteus,* some of which can cause mild to serious illness depending on species, strain, whether the host is immune-compromised, and location in the body.

- ⊙ **Pathogens** – This would include species within the genera *Salmonella, Serratia, Shigella,* and the world-famous *Yersinia.* These genera have more than a few organisms that can be fairly virulent, successfully attacking even hosts that are otherwise in good health.

Like many of the assays that you have completed thus far, the constituent tests of the IMViC series operate on the principle of identifying the activity of certain biochemical pathways or whether certain enzymatic reactions are taking place.

- ⊙ **Indole** – This is an assay that will help you know something about the type of metabolism that microorganisms engage in. Specifically, not all microbes can metabolize the amino acid **tryptophan.** Some organisms express the **trytophanase** enzyme which is an initial step of the metabolism of this amino acid. One of the products of this hydrolysis is indole, which can be detected by adding Kovac's reagent to

SIM agar which has been inoculated and incubated. Kovac's reagent is composed of butanol, hydrochloric acid, and *p*-dimethylaminobenzaldehyde which will complex with indole yielding a cherry/rose color.

- **Methyl red** – Glucose (dextrose) is a substrate that is used by all Enterobacteriacaeae and a large majority of all other microbes as well. Not all organisms use the same dextrose metabolism pathways and so the presence or absence of these pathways is another avenue that we can use to differentiate them. The MR test will allow you to determine whether an organism will ferment dextrose while producing *stable* acid end products. Some organisms like *E. coli* or *S. aureus* will create acid end products that remain acidic as they do dextrose fermentation resulting in a pH of ~4. You will simply add a few drops of the pH indicator methyl red after incubation. Presence of acid will cause the indicator to turn the broth red. If the broth does not change color or becomes slightly yellow after addition, you have little or no acid present.

- **Voges-Proskauer** – The same broth is used in this assay as was used in the MR test. Some microorganisms such as *K. aerogenes* initially produce acidic end products but later enzymatically convert them to non-acid compounds which results in a slightly higher pH (~6) at the end of the assay. This test will allow you to determine whether any of those non-acid end products of dextrose fermentation are present. This assay requires Barritt's reagents A & B which are alcoholic α-naphthol and 40% KOH, respectively. It takes time for the chemistry to take place, so after incubation, add your reagents then observe the result at 15 or 20 minutes to see whether a rosey red color has developed on the surface of the assay. If you have no color change within 20 min, it means your organism doesn't produce non-acid end products during dextrose fermentation.

- **Citrate** – As you have seen, microorganisms exhibit an array of fermentation pathways and substrates. Some have the ability to ferment citrate if no other fermentable substrates (like dextrose or lactose) are present. To do this, microorganisms must first transport the citrate into their cells via **citrate permease.** Once in the cells, citrate is cloven into oxaloacetic acid and acetate via **citrase.** These products can then be converted to pyruvate and CO_2. If we grow our microbes on Simmons' citrate agar, these metabolic reactions will cause the medium to become alkaline as the CO_2 reacts with Na in the medium to become sodium carbonate (Na_2CO_3). This compound is alkaline and will cause the bromthymol blue pH indicator to change from green to a deep blue.

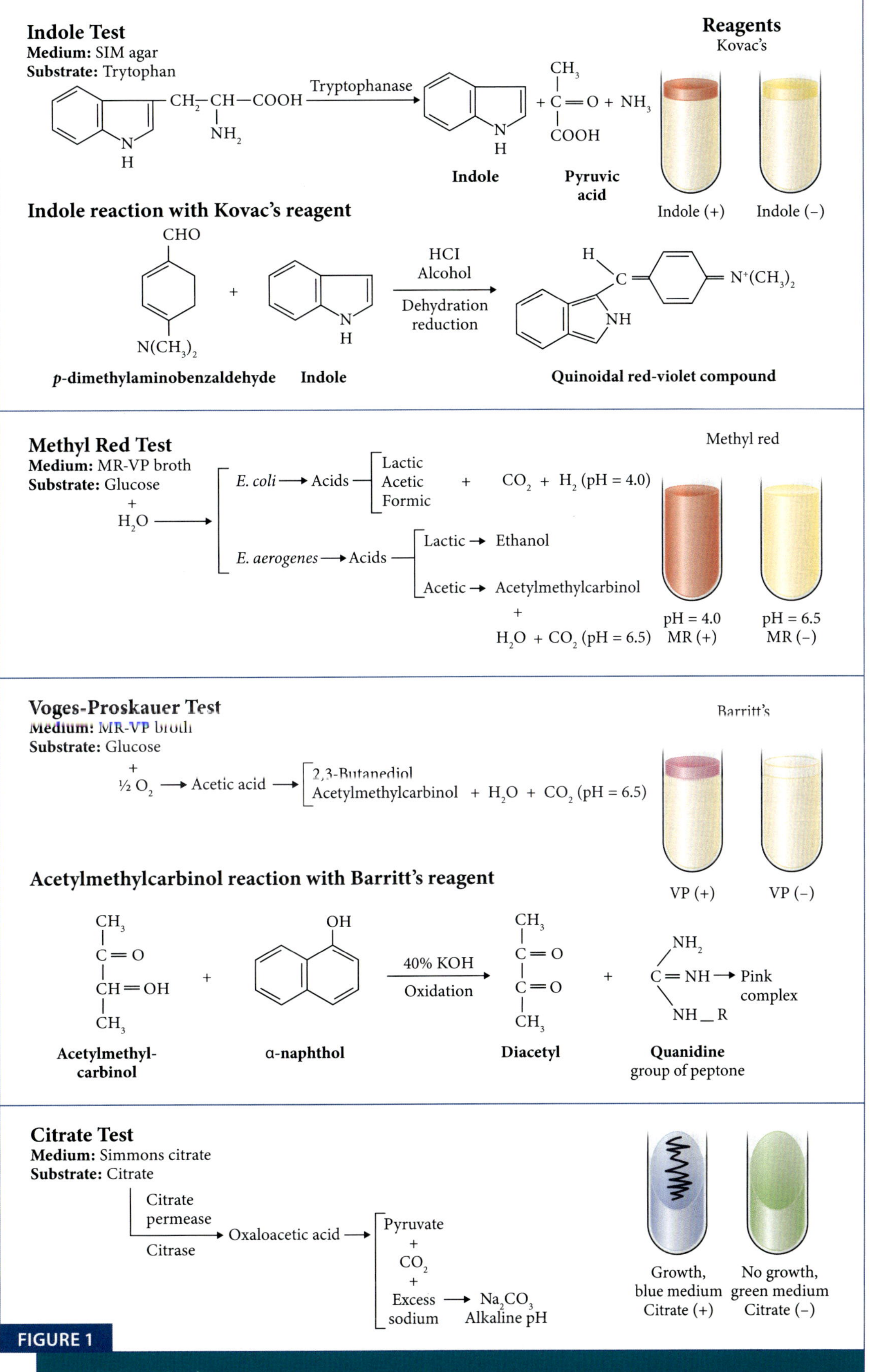

FIGURE 1

IMViC Test Series

PROCEDURE

1. Obtain your maintenance culture:

    a. A single MR/VP tube and an empty, sterilized tube.

    b. A Simmons citrate agar slant.

    c. An EnteroPluri-Test.

2. Inoculate the broth in the standard fashion.

3. Inoculate the EntroPluri-Test according to the manufacturer's instructions.

4. Incubate all media at 37° C for 24 hrs.

5. Turn to the 24 Hour Return section of this manual and follow the directions there.

    a. Be sure to observe the EnteroPluri-Test within the specified time frame. Premature or delayed observations will invalidate results.

NITRATE REDUCTION
A TYPE OF ANAEROBIC RESPIRATION

Fermentations aren't the only type of metabolism that we can use to classify organisms. There is more than one type of respiration, for instance. Organisms might be aerobic or they might be anaerobic or they may be aerotolerant. The types of respiration you find operating in organisms within those categories are different and may therefore be useful for classification. In this laboratory exercise, you will be investigating the ability of some organisms to use nitrate (NO_3^-) as a source of terminal-electron-accepting oxygen.

Nitrate reduction has more than one step, and not all organisms do all the steps. Hence, there are partial and complete NO_3^- reductions. For the purposes of this lab, we will define them as follows:

Partial – Ionic product that stays in solution

$$NO_3^- + 2\ H^+ + 2\ e^- \rightarrow NO_2^- + H_2O \text{ (Nitrate reductase)}$$

Complete – Gaseous product that escapes the solution

$$NO_2^- + 2\ H^+ + e^- \rightarrow NO + H_2O \text{ (Nitrite reductase)}$$
$$2\ NO + 2\ H^+ + 2\ e^- \rightarrow N_2O + H_2O \text{ (Nitric oxide reductase)}$$
$$N_2O + 2\ H^+ + 2\ e^- \rightarrow N_2 + H_2O \text{ (Nitrous oxide reductase)}$$

It isn't uncommon for microorganisms to have both aerobic and anaerobic respiration pathways and to use whichever one is more apt for the given circumstances. All the organisms that we use in this lab are capable of growth in the presence of atmospheric oxygen (O_2), but they may be **obligate aerobes, facultative aerobes,** or **aerotolerant** (refer to Catalase Test Lab).

This matters because among the organisms we use in this lab, it is only the facultative anaerobes who may do NO_3^- reduction. Facultative anaerobes will preferentially use O_2 via their aerobic pathway if O_2 is available. By growing our microbes in **trypticase nitrate broth** that has just enough agar added to it to make it semi-solid, we slow the diffusion of O_2 into the medium. This means that any organisms with an aerobic pathway will use up the O_2 at a faster rate than it can diffuse back in. This will create an anaerobic environment extending from just below the medium surface to the bottom of the tube. In response, any facultative anaerobes will change from their aerobic respiration to anaerobic respiration. Once that change-over occurs, any organisms that can reduce NO_3^- will begin to do so.

We will use two (possibly three) reagents to help us determine whether our organisms are capable of doing nitrate reduction. **Reagent A** is **sulfanilic acid** and **Reagent B** is **α-napthylamine,** and depending on your initial result, you may need to use **zinc powder.** Reagents A and B are colorless liquids that together will react with any nitrous acid (HNO_2) present in the medium to form a deep red color. Zinc is capable of catalyzing the reduction of NO_3^- to NO_2^- all by itself.

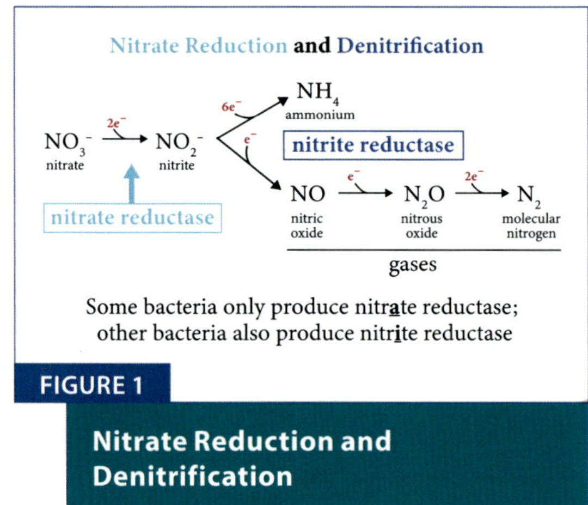

FIGURE 1

Nitrate Reduction and Denitrification

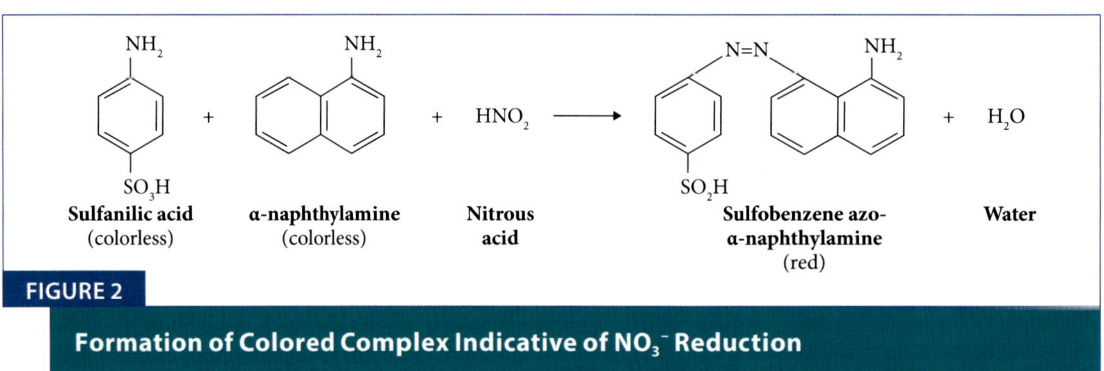

FIGURE 2

Formation of Colored Complex Indicative of NO_3^- Reduction

PROCEDURE

1. Obtain a tube of trypticase nitrate broth for every organism you are to investigate.

2. Inoculate your media with those organisms by transferring one loop of the organism into the broth.

3. Incubate at 37° C for 24 hrs.

4. After incubation, add 5 drops of Reagent A followed by 5 drops of Reagent B.

5. If the reagents bring about a red color within 30 seconds, your experiment is done.

6. If the reagents bring about no color change, add a small scoop of zinc powder and watch for any development of a red color.

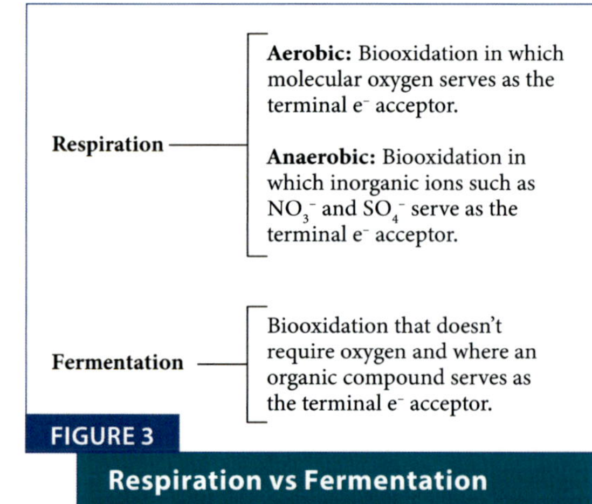

FIGURE 3

Respiration vs Fermentation

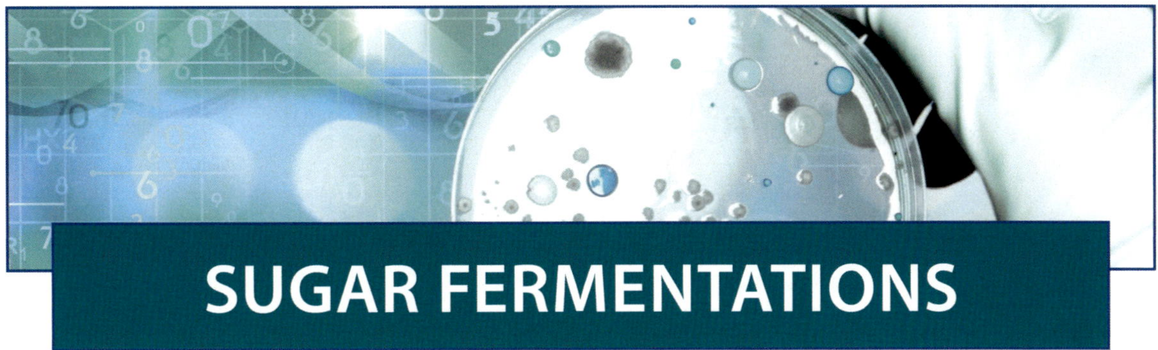

SUGAR FERMENTATIONS

As we've discussed before, fermentations can be diagnostic for identifying microorganisms. In this exercise, you will determine whether some microorganisms can or cannot ferment two carbohydrates: dextrose and sucrose.

The ability of organisms to utilize **catabolic compounds** is determined by that organism's complement of enzymes. A very common catabolic pathway is **glycolysis** whose end products still contain a significant amount of energy. Some microorganisms are capable of fermenting these glycolytic end products to keep their co-enzymes (like NAD+) free to continue facilitating more glycolysis. Microorganisms may do this using an aerobic pathway, others use an anaerobic pathway, and still others may use either depending on the situation, and some organisms do not do this at all. Each pathway has specific products, some of which, such as acids and gasses, are readily diagnostic.

In this exercise, you will be using standard test tubes containing a sugar-fortified nutrient broth, a pH indicator, and a Durham tube. The nutrient broths will contain either dextrose or sucrose along with phenol red which has a red/purple color at a neutral pH and turns yellow in the presence of acid. The **Durham tube** is simply a small test tube that has been placed inverted in the broths. If any of the organisms ferment glycolytic products and thereby produce one or more gasses, the Durham tube will capture a fraction of the gasses forming a gas bubble.

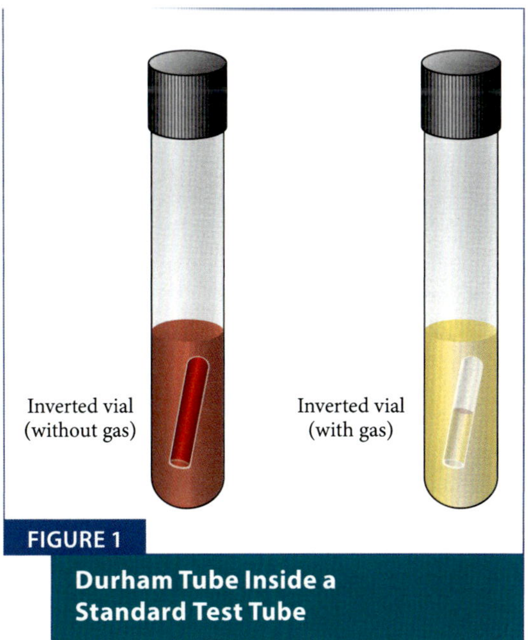

FIGURE 1

Durham Tube Inside a Standard Test Tube

Be advised that an absence of detectable fermentation doesn't mean the organism didn't grow in the medium. There are other energy-containing compounds in the nutrient mix such as peptones, the processing of which produces ammonium hydroxide (NH_4OH) which is alkaline.

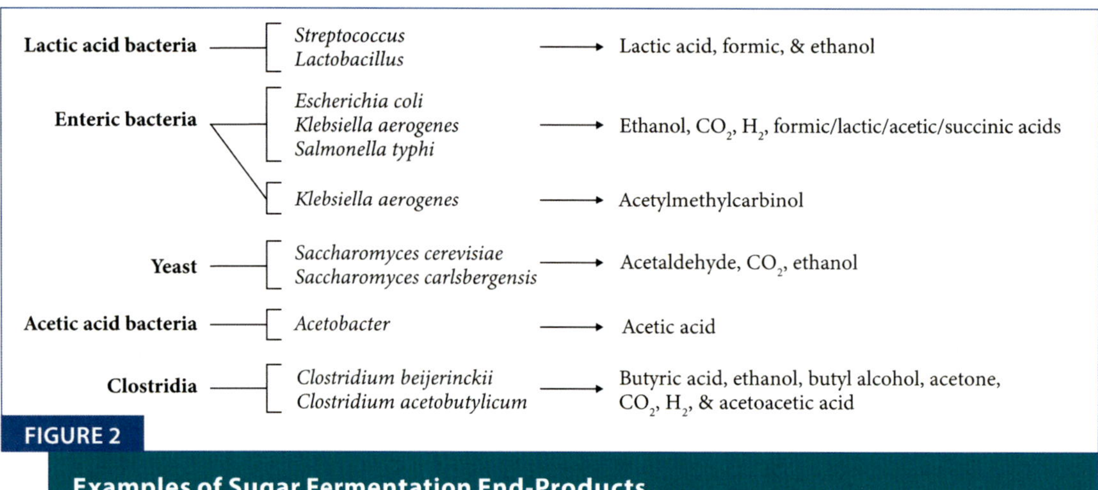

FIGURE 2

Examples of Sugar Fermentation End-Products

PROCEDURE

1. Obtain one of each broth for every organism you are investigating.

2. Inoculate the broths with a single loop of microbial culture. Be careful not to stir up the broth as you inoculate as you may introduce air bubbles into the Durham tube.

3. Incubate at 37° C for 24 hrs.

4. After incubation, observe the color of the broths and whether the Durham tube captured any gasses. You should see no gas production in tubes that have not acidified.

OXIDASE TEST

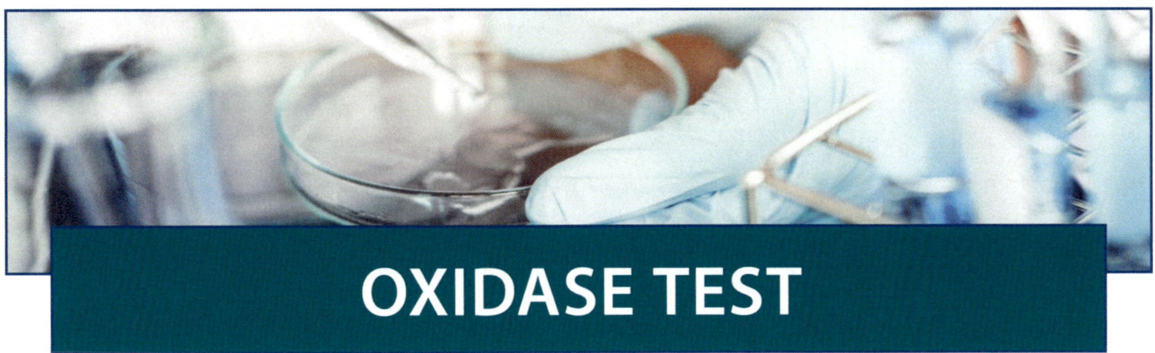

Cytochromes are iron-containing heme-proteins that are critical to **electron transport chains (ETCs).** Various micro-organisms may use various ETCs within which are various cytochrome compounds which function as e⁻ carriers. To be effective e⁻ carriers, cytochromes must be repeatedly reduced and then oxidized. Interruption of these reduction/oxidation loops is a type of poisoning which can lead to organism death. The oxidation part of the loop is accomplished via **cytochrome oxidase enzymes.**

As you may deduce from its name, cytochrome oxidase catalyzes the oxidation of a reduced cytochrome using molecular oxygen (O_2). This oxidation produces H_2O and hydrogen peroxide (H_2O_2). Aerobic bacteria, some facultative anaerobes, and some microaerophiles exhibit cytochrome oxidase activity. The ability isn't universal and can therefore be used for microbial classification. For instance, it is a fairly quick method of differentiating between the small, gram– bacilli of the Enterobacteriaceae (oxidase negative) and the small, gram– bacilli of the genera *Neisseria* and *Pseudomonas* (oxidase positive).

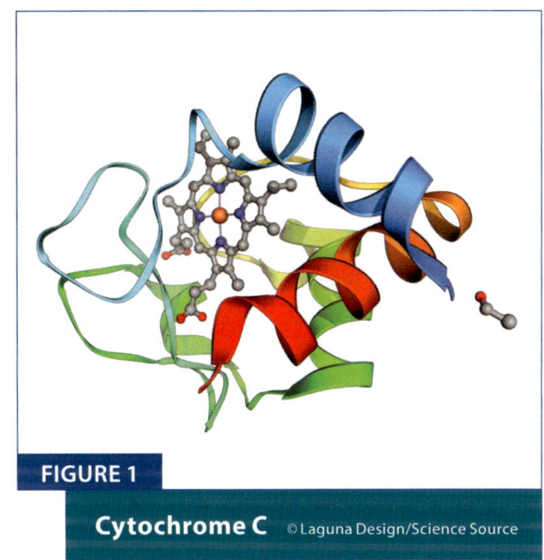

FIGURE 1

Cytochrome C   © Laguna Design/Science Source

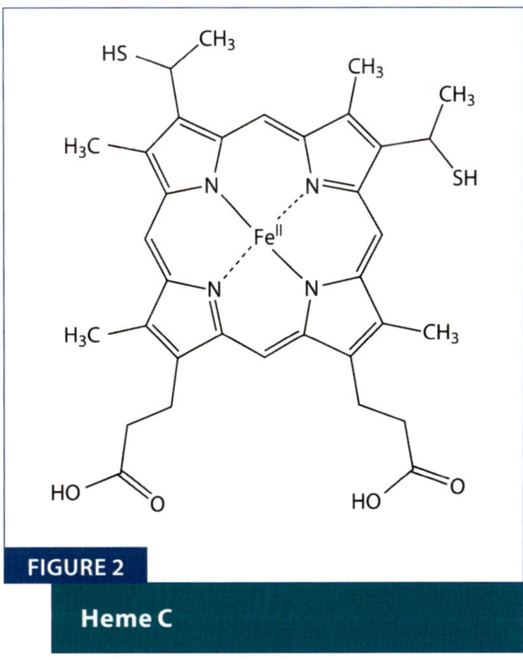

FIGURE 2

Heme C

The expression of cytochrome oxidase can be detected via the addition of tetramethyl-p-phenylenediamine which is initially colorless but will turn a deep blue-black in the presence of cytochrome oxidase. This is the method you will use to classify bacteria during this laboratory exercise.

PROCEDURE

Obtain plates of the organisms you are to investigate and a dropper bottle of tetramethyl-p-phenylenediamine.

DIRECT METHOD

1. Place **one drop** of the reagent directly onto a single colony in your plate, ***being careful not to touch the nozzle of the bottle to the colony***. If you touch the colony, you will contaminate the bottle and ruin the rest of the reagent.

2. Watch for a change of the reagent from colorless to blue-black (usually within 10 seconds).

3. This method should not be used with colonies on blood agar, as the agar is likely to contain active oxidizers within it, which can yield false-positive results. For colonies grown on blood agar, use the filter paper method.

FILTER PAPER METHOD

1. Obtain sterilized filter papers.

2. Place **one drop** of tetramethyl-p-phenylenediamine on a filter paper.

3. Using your inoculating loop, transfer a colony from your plate to the reagent-wet filter paper.

4. Watch for a color change to blue–black within 10 seconds.

If you use the filter paper method, beware of using reactive wire loops such as NiChrome. Generally, the use of a loop constructed of anything other than an inert metal such as platinum (Pt) will cause spontaneous oxidation of the reagent. Thankfully, this loop-induced oxidation is typically slower than metabolic oxidation, taking place after ~10 seconds.

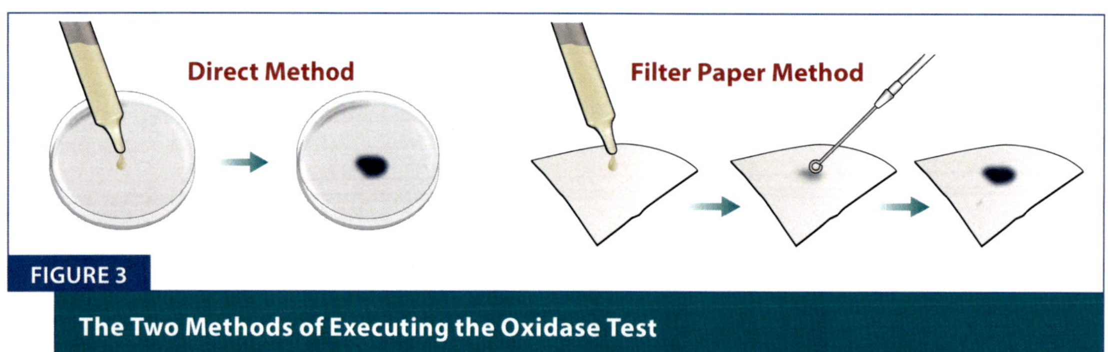

FIGURE 3

The Two Methods of Executing the Oxidase Test

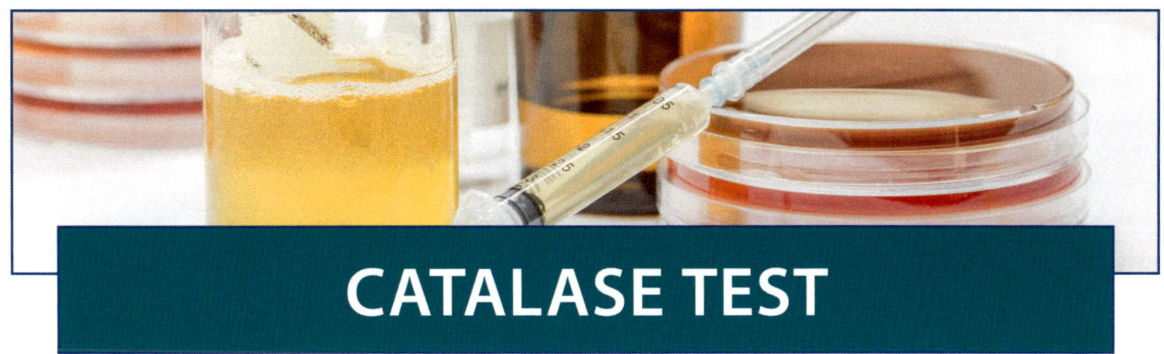

CATALASE TEST

Not all the byproducts of aerobic metabolism are harmless to the cell. Some can be extremely harmful, such as highly-reactive hydrogen peroxide (H_2O_2) or superoxide (O_2^-). If not converted to something less reactive, these strong oxidizers will rampage around inside the cell oxidizing proteins, enzymes, DNA, RNA, and all manner of things. O_2^- is so toxic that it is used by the human immune system as a means of killing microbial invaders. It is therefore detrimental for a microbial cell to generate these compounds as part of their standard metabolism and to have no means of rendering them less harmful. These methods of catalytically converting toxic compounds to less-harmful by-products are not universal, so they can be used to classify microorganisms.

In this laboratory exercise, you will be investigating whether various microorganisms express **catalase** which catalyzes the breakdown of $2H_2O_2 \rightarrow 2H_2O + O_2$. Not all aerobic organisms express catalase, and those that don't do still express **superoxide dismutase** which catalyzes $O_2^- \rightarrow H_2O_2$ which, while still toxic, is much less toxic than superoxide. Anaerobic organisms do not express catalase, peroxidase, or superoxide dismutase and therefore cannot survive in environments with significant amounts of O_2.

PROCEDURE

1. Obtain the plates of your given microorganisms, a bottle of diluted H_2O_2, and a dropper.

2. Place ***one drop*** of the dilute H_2O_2 directly onto a colony on your plate.

3. Watch for immediate bubbling and frothing of the liquid which indicates a positive test.

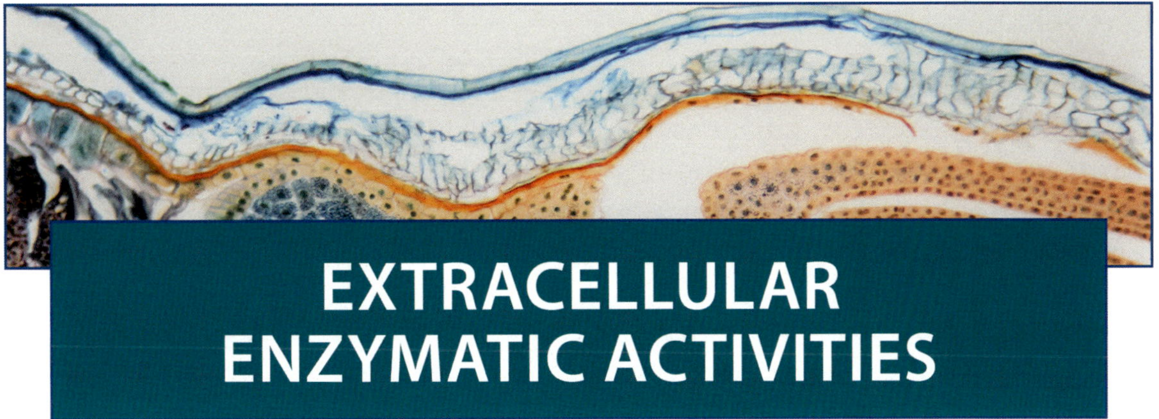

EXTRACELLULAR ENZYMATIC ACTIVITIES

Some useful molecules are too big to bring into the cell. There are plenty of high-molecular-weight molecules that could be useful, if only a cell could manage to get them imported. Polysaccharides, lipids, and proteins all fit this description. One way that a microbial cell may tap into these resources is to produce and secrete **extracellular enzymes** or **exoenzymes.**

Exoenzymes are produced by many organisms, usually gram+ bacilli. Most of these enzymes are **hydrolases** that use water in the process of splitting the bonds that tie together large macromolecules.

In this laboratory exercise, you will investigate whether several microorganisms produce exoenzymes capable of hydrolyzing a few macromolecules:

STARCH

This is a branching polymeric carbohydrate (polysaccharide) that consists of a large number (poly–) of glucose molecules (–mers) covalently bonded together by **glycosidic bonds.** These bonds can be hydrolyzed by **amylases,** which split the large starch molecules into smaller chunks. The smaller chunks are **dextrins** and **maltose.** Dextrins will form spontaneously if you expose starch to high heat and are partially responsible for the browning of toast. Dextrins are split until you are left with molecules of **maltose,** which consist of two glucose molecules bound together. Finally, maltose is hydrolyzed by **maltase** which yields soluble glucose that can be brought into the cell.

You will be using agar that includes 8% starch by dry weight. This type of agar is a translucent white, and it often continues to look that way even if you grow an organism that expresses amylases. When you expose starch to a tincture of iodine, the iodine will temporarily stain the starch black. Hence, you will use a tincture of iodine to show you whether any of the organisms hydrolyzed the starch. If any of them do, the media immediately adjacent to the microbial growth will remain translucent in the presence of the iodine while the medium further away will turn black.

GELATIN

Gelatin is an incomplete protein because it lacks tryoptophan, but that doesn't stop microorganisms from using it as a nutrient. Proteins are polymers of amino acids where the amino acids are covalently bonded by **peptide bonds.** Gelatin is formed via hydrolysis of collagen. Gelatin is solid below 25° C, and above that, it will liquefy. Some organisms are capable of producing **gelatinase** which will hydrolyze gelatin. That hydrolysis causes the medium to liquefy, and it will remain liquid at room temperature and even down to temperatures near the freezing point of water. The value of gelatin for identifying bacteria has been recognized since the late 1800's.

You will be using TSB into which 12% gelatin by dry weight has been dissolved. You will stab inoculate the medium in the test tubes and incubate it for 48 hours then place it in the refrigerator until it fully cools down. Any tube that remains liquid after refrigeration indicates that the organism did **rapid gelatin hydrolysis.** Any that are solid may be re-incubated for 5 days and then refrigerated. Tubes that are liquefied after the second incubation indicate that the organism is capable of **slow gelatin hydrolysis.**

CASEIN

This is one of the components that make milk white, and it is the major protein in milk which is why it is otherwise known as "milk protein." Proteins must be hydrolyzed before they can be brought into a cell, and the hydrolysis proceeds in steps: protein -> peptones -> polypeptides -> dipeptides -> amino acids. Collectively, these steps are **proteolysis,** and it is done by **proteases.**

You will be using a nutrient agar which includes 70% by dry weight instant non-fat dry milk. This agar typically has a milk-like color that is more opaque than starch agar. In this preparation, which excludes milk fats, it is the casein that gives the agar its milky color. You will inoculate your casein agar plate and incubate it as normal. After incubation, if you find that the medium immediately under and adjacent to the microbial growth has become transparent, that indicates that the organism on that plate produced a **caseinase** that hydrolyzed the casein in the medium.

LIPIDS

Lipids, also known as fats, are large macromolecular compounds that are often used to store a tremendous amount of energy. Lipids are a class of compounds encompassing a number of different basic structures. One common subclass of lipids is **triglycerides,** which consist of a glycerol cap attached to three fatty acid chains. These are initially hydrolyzed by a class of **lipases** called **esterases** which split the fatty acid tails off the glycerol cap. Once split off, the fatty acids and the glycerol can be brought into the cell and further broken down so as to be run through the Krebs Cycle.

You will be using a nutrient agar that includes 30% tributyrin by dry weight. The presence of the high concentration of lipid causes the otherwise clear agar to become somewhat cloudy. After inoculating and incubating in the standard way, if you find that the medium immediately under and adjacent to the microbial growth has become transparent, that indicates that the organism in that plate produced a lipase capable of hydrolyzing the **ester** bonds binding the fatty acids to the glycerol.

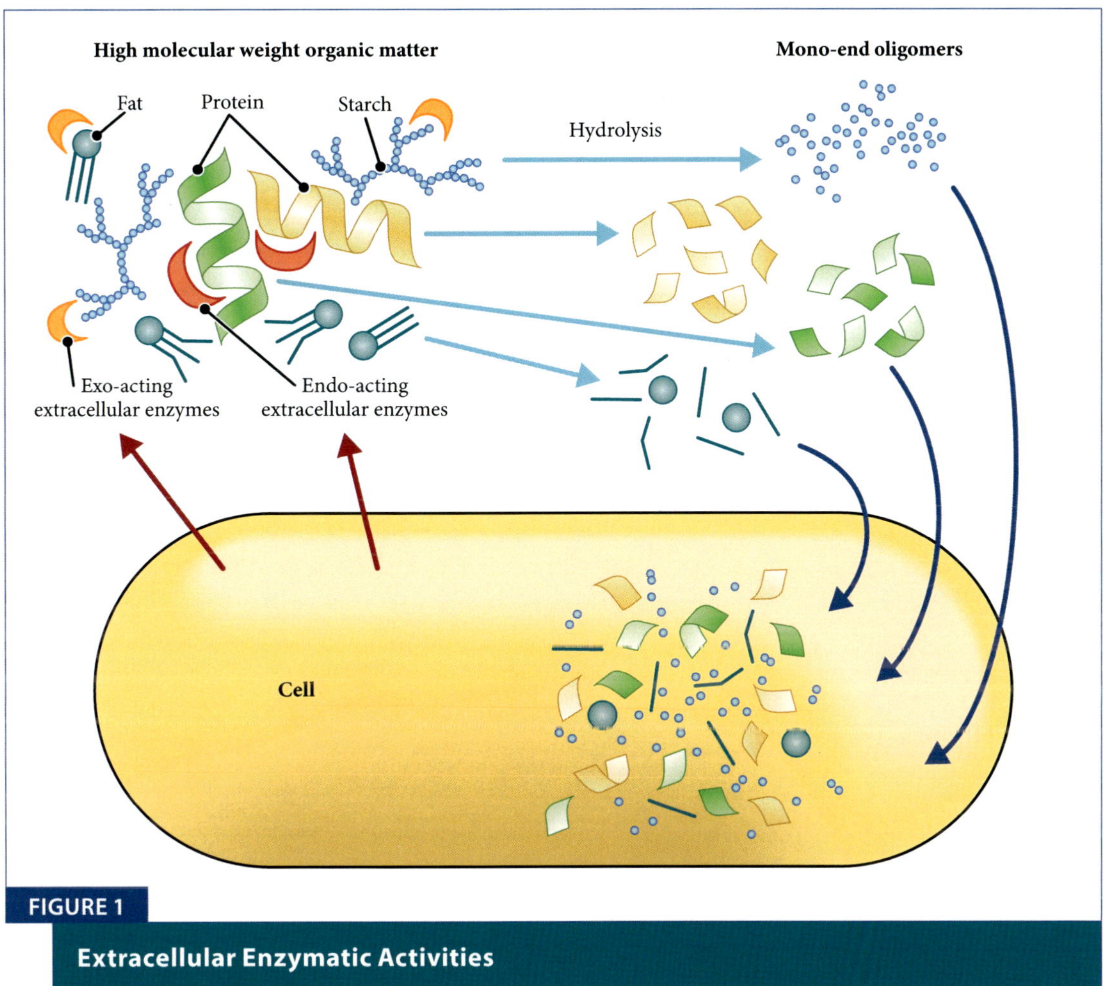

FIGURE 1

Extracellular Enzymatic Activities

PROCEDURE

1. Obtain your maintenance culture and a medium of each type.

2. Inoculate your organism onto each agar plate with a standard 3-way streak method.

3. Inoculate your organism into the gelatin tube by stabbing once into the center of the gelatin surface.

4. Incubate the agar plates at 37° C for 24 hrs.

5. Incubate the gelatin tube at 37° C for 48 hrs (rapid method).

DIFFERENTIAL, SELECTIVE & ENRICHED MEDIA

All of the various microbiological growth media that we have used this semester have been **nutrient media.** That is, they have the basic nutrients that at least *some* microbes need for growth and reproduction. Some of the media that we have used are more than just nutrient media, though. These special-purpose media have additional ingredients or perhaps lack specific components. Some of these media are **selective media** and some of them are **differential media** and some do both jobs at once.

Selective media are designed to inhibit the growth of some types of microorganisms while allowing the growth of other types. These "types" may encompass categories like bacteria vs. fungi (TSA vs. Sabouraud agar) or gram+ vs. gram– (phenyethyl alcohol agar vs. MacConkey agar). In this laboratory course, you will use one selective medium:

- **Phenylethyl alcohol agar (PeOH)** – Contains 6% phenylethyl alcohol (a.k.a **phenethyl alcohol**) which is somewhat inhibitory to most gram   organisms and to a few gram+ organisms. This medium is used to isolate gram+ organisms, especially if your culture has become contaminated with swarming *Proteus* species. If an organism's survival and reproduction are inhibited, then its colonies and colony morphology should be smaller and thinner than they would be on a non-selective medium.

Differential media are designed to change in color or clarity in response to certain types of microbial biochemistry. The ability to ferment certain carbohydrates and produce acidic end products is a common analytical category. Change out the carbohydrates or the pH indicator and you have a different type of differential medium. Add in some ingredients that inhibit certain microorganisms and you have a medium that is both differential and selective. In this laboratory course, you will use the following media that are simultaneously differential and selective:

⊙ **Mannitol salt agar (MSA)** – This is a medium that it commonly used in clinical settings to isolate *Staphylococcus* species. It is able to do this because the medium contains 7.5%–10% NaCl which is a relatively high concentration of salt that inhibits the growth of most clinical bacteria outside of genus *Staphylococcus*. Moreover, it can be used to rapidly identify which are likely *Staphylococcus aureus* because the medium contains the carbohydrate mannitol and the pH indicator phenol red which together allow the differentiation of colonies capable of fermenting mannitol (common in *S. aureus*) and those that are not. The medium is initially red, and if the organism produces acid (the byproduct of mannitol fermentation), the medium will turn yellow.

⊙ **MacConkey agar (MCA)** – This medium allows for the isolation of gram– organisms and for the differentiation between those that can ferment lactose while producing acid and those that cannot. Organisms that are gram– and ferment lactose are presumed to be **coliform bacteria.** Gram– organisms that do not ferment lactose are presumed to be **typhoid, paratyphoid,** or **dysentery bacteria.** The medium contains the dye crystal violet and bile salts which are inhibitory to gram+ organisms and fungi. Also present are the carbohydrate lactose and pH indicator neutral red which together allow an investigator to determine whether there are organisms present capable of fermenting lactose and producing an acid as a byproduct. The medium is initially a light pink/red color and will remain so if gram– organisms incapable of fermenting lactose are grown on it; the colonies will generally be transparent and the same color as the medium. If an organism is

FIGURE 1

MSA. After 24 hours, this inoculated Mannitol Salt Agar (MSA) culture plate cultivated colonial growth of: 1) *Micrococcus* sp; 2) *S. epidermidis;* and 3) *S. aureus.*   By Navaho [CC BY-SA 4.0]
https://commons.wikimedia.org/wiki/File:Chapmanes.jpg

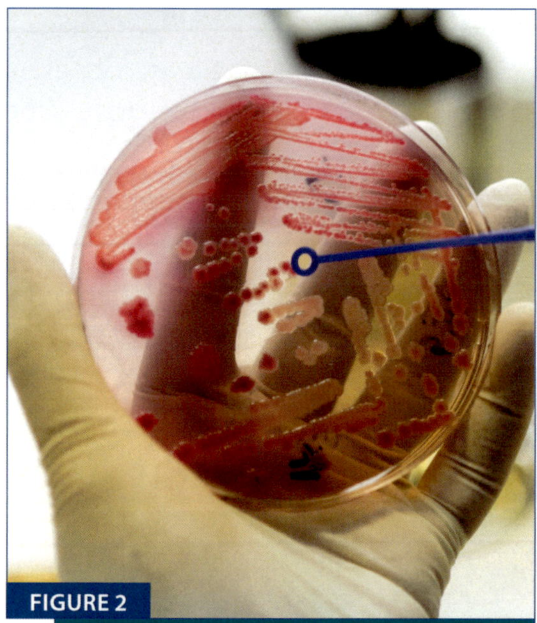

FIGURE 2

MCA. This is what 18 hrs of growth looks like on an MCA plate inoculated with *E. coli.*

capable of lactose fermentation, then the colonies will turn an opaque, bright pink color. If the organism vigorously ferments the lactose, producing copious amounts of acid end products, the medium immediately adjacent to the colonies will lose clarity and become more opaque as the bile salts precipitate out of the acidified medium and form crystals within the agar matrix.

- ⊙ **Eosin-methylene blue agar (EMB)** – This medium shares a number of similarities to MCA in that it also isolates gram– organisms and allows the investigator to determine if any ferment lactose producing acid. The medium contains the dyes eosin Y and methylene blue which both inhibit gram+ bacteria and will be absorbed by bacterial colonies under acidic conditions. These dyes are not as strongly inhibitory as the combination of CV and bile salts in MCA, so some small growth of gram+ organisms is common on this medium. Also present in the medium is lactose, which may or may not be fermented by the organisms grown on the medium. The medium is initially a transparent purple color, and organisms incapable of lactose fermentation won't change this; they'll be transparent. If an organism is capable of fermenting lactose, then at least the center of the colony will become more opaque as the acidic environment encourages the colony to absorb the dyes. It is common for not just the center but the entire colony to become a dark, opaque blue–black. This medium can be useful in specifically identifying *Escherichia coli* and *Klebsiella aerogenes* whose colonies will be metallic green and a slimy pink color respectively.

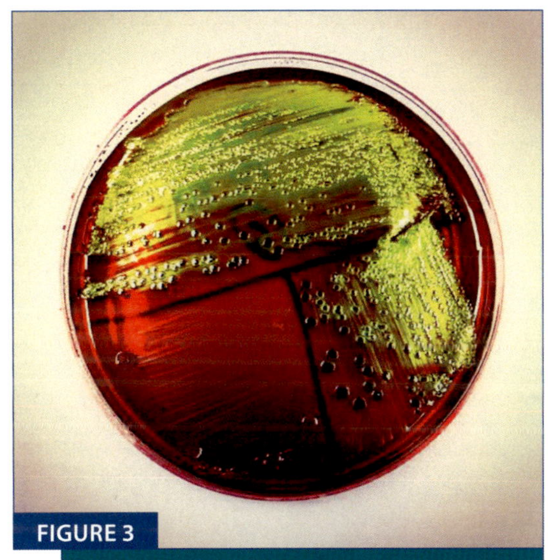

FIGURE 3

EMB. *E. coli* culture on EMB agar, a selective and differential medium in which the growth of gram-positive bacteria is inhibited. Moreover, those organism who can ferment lactose, such as *E. coli*, will absorb the dye contained on the medium and will have a distinctive metallic green as shown in the picture.

By Carmen Moreno González [CC BY-SA 4.0]
https://commons.wikimedia.org/wiki/File:E.coli-fields.JPG

- **Sheep blood agar** – This is technically an **enriched medium** because blood is a very high-nutrient ingredient and contains many unique trace components. Media of this kind permit the growth of **fastidious organisms.** This agar can also function as a differential medium. Specifically, this medium can help an investigator determine whether any of their organisms are capable of performing **hemolysis,** which is the breakdown of red blood cells (RBCs) and their hemoglobin. Hemolytic activity is a **pathogenic indicator** which is caused by the **virulence factor** hemolysin (an enzyme). Therefore, hemolytic activity can be a clinically important factor to identify. Beware, however, that some species, such as some of the *Streptococci,* express **hemolysins** that are deactivated by the presence of O_2; therefore it is common practice to stab the blood agar plate in an open spot with your loop after inoculating it. The lower-O_2 concentration within the agar may reveal the presence of these hemolysins. There are three classes of hemolytic activity:

  - **α (alpha)** – Incomplete hemolysis where hemoglobin is partially metabolized to methemoglobin (pronounced "met-hemoglobin") which gives the immediately adjacent medium a green/grayish tinge. Some α hemolyzers will clear the medium under the colonies but retain the green/gray color.

  - **β (beta)** – Complete hemolysis wherein the RBCs are completely lysed and the hemoglobin is more fully metabolized. Organisms capable of this feat will have a zone of transparent medium under and immediately adjacent to colony growth. This transparent zone will also be much paler, no longer red, and will have no green/gray coloration in the medium.

  - **γ (gamma)** – No hemolysis. Organisms that exhibit this class of hemolysis are actually not doing hemolysis at all. Colonies will commonly appear a white/gray color and will affect no change in medium color or clarity.

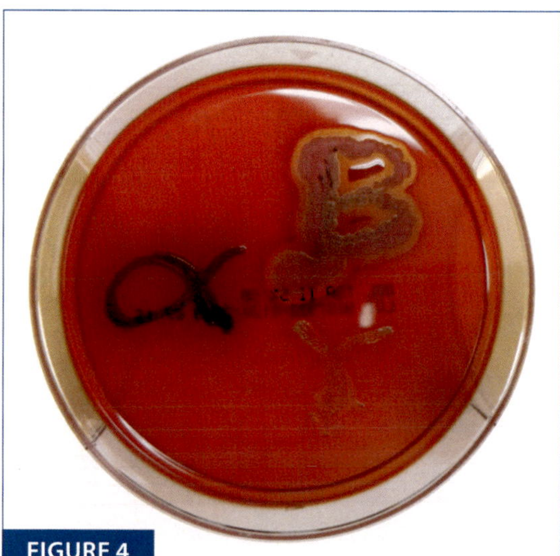

FIGURE 4

SBA. Petri dish containing sheep's blood agar (SBA) inoculated with *Escherichia coli* (alpha), *Bacillus cereus* (beta), and *Enterococcus faecalis* (gamma). This is the result after 48 hrs of incubation at 37°C. Note the green color of the colonies and in the medium surrounding the alpha, the clear zone around the beta, and the fact that gamma changed nothing about the medium.

PROCEDURE

1. Obtain your maintenance culture and a plate of each medium.

2. Inoculate your organism onto the surface of the medium in each plate.

3. Incubate all plates at 37° C for 24 hrs, except for the blood agar plates which should be incubated for 48 hrs.

4. After incubation, observe your plate and interpret any changes in the medium or colony appearance according to the principles laid out above.

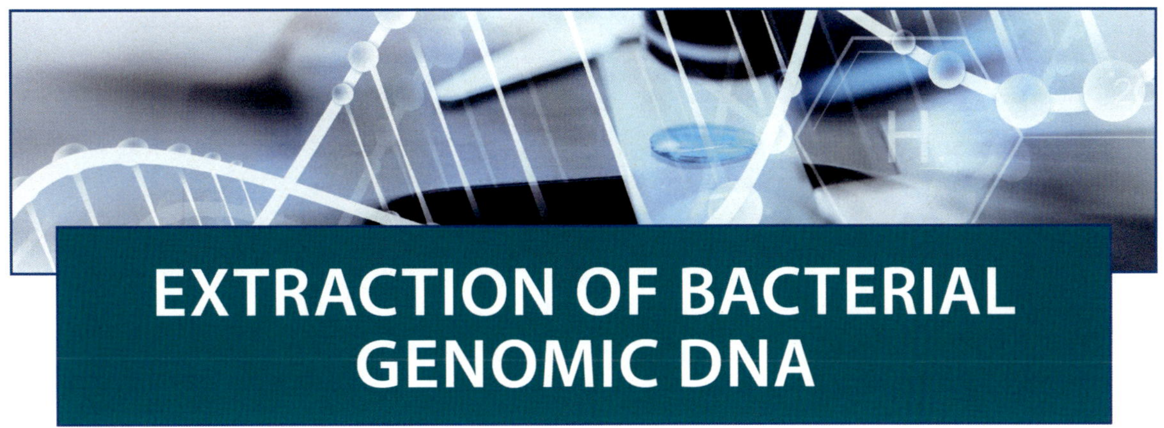

EXTRACTION OF BACTERIAL GENOMIC DNA

DNA extraction, polymerase chain reaction (PCR), and genetic sequencing are a revolution akin to the invention of the microscope. We've learned enough about the function and makeup of DNA and RNA over that last few decades that we can now manipulate genomic machinery to expedite medical and environmental advances. As biological science majors in the 21st century, you should be familiar with these tools.

Over the next 3 lab sessions, we will go through the following steps of:

⊙ Extracting and purifying bacterial genomic DNA.

⊙ Amplifying a target sequence and confirming that our amplification actually produced a product.

⊙ Sending our amplified sequence out to be sequenced and inputting the resulting sequence data into bioinformatics software to determine how homologous your sequence is to sequence data deposited in the NCBI GenBank database.

PART 1: DNA EXTRACTION & PURIFICATION

To fry an egg, first you must crack the shell. Similarly, if the chromosomal DNA is locked up inside of bacterial cells, we cannot retrieve it for our analytical purposes. Therefore, we must first get the DNA out of the bacterial cells. After removing the DNA from the bacterial cells, we need to remove potentially inhibitory proteins as well as other cellular debris. This is done in the following stages:

⊙ **Harvest cells** – Here, you will first use a centrifuge to separate particles based on mass and shape, concentrating the bacterial cells into a formed pellet at the bottom of the microfuge tube.

⊙ **Resuspend cells in a buffer** – The pelleted cells are now clumped together, but the DNA in each cell is not readily accessible, so the pelleted cells need to be separated and resuspended in a buffer that will stabilize the DNA once it is outside the cell.

- ⊙ **Eliminate protein contaminants** – Here you add Proteinase K, a serine protease that hydrolyzes a variety of peptide bonds, to your suspended cells and then heat the cells up to the temperature at which the enzyme is active (~50°C). The proteinase will chew up any external protein contaminants and, when the cell is cracked open, do the same to the interior ones as well.

- ⊙ **Lyse cells** – This is where you split the bacterium open by adding a lysis buffer to the cell solution. Lysis buffers typically include things like a buffer (monosodium phosphate, Tris-HCL, etc.), salts (NaCl, KCl, etc.), and a detergent which will disrupt cell membranes. Since the cells walls of gram+ bacteria possess a thick layer of peptidoglycan, the addition of lysozyme is required when lysing gram+ cells to increase the efficiency of the DNA extraction.

- ⊙ **DNA purification** – Once the cells are lysed, the DNA is floating in a pool of contaminants which can interfere with the gene amplification that we are planning to do. You can remove these contaminants by causing the DNA to stick to something and then wash away all the contaminants that do not stick. This is accomplished using a silica membrane (column), salts that will cause the DNA to bind to the silica, and wash solutions that will wash away contaminants without knocking the DNA off the column. Be sure to spin your column dry of ethanol (EtOH) at the end of your washes to prevent DNA damage.

- ⊙ **DNA elution** – Your DNA is adhered to the silica column, and you need to release the DNA into the liquid suspension . You will rehydrate and release the DNA off the column and into another buffer (often Tris at pH 8-9). From here, you can use your clean DNA for whatever manipulation you have planned.

PART 2: AMPLIFYING THE 16S rRNA GENE & CONFIRMING PRODUCT

In this portion of the project, you will use the DNA that you extracted and purified last week as a template for PCR. PCR, or **polymerase chain reaction,** is a method of performing *in vitro* DNA synthesis. For our project, we want to use PCR to amplify a specific target sequence within the bacterial genomic DNA you extracted. We want to use our PCR product to help us identify the organism. We will therefore be amplifying the 16S rRNA gene of your bacterial organism, which can efficiently identify and categorize bacterial organisms to a particular genus and to a lesser extent a particular species.

16S rRNA is the RNA component of the 30S subunit of the prokaryotic ribosome. The utility of this particular gene for constructing prokaryotic phylogenies has been recognized since at least the late 1970's (Woese & Fox, 1977). This gene is highly conserved, yet it has a mix of variable and highly conserved regions. There have been thousands of 16S rRNA gene sequences uploaded to sequencing databases over the decades, making this method the gold standard for general bacterial identification.

MASTER MIX COMPONENTS

⊙ **Autoclaved deionized (DI) water** – The universal solvent that we will use to hold all of the following components together through the PCR process.

⊙ **Buffer** – These usually consistent of a Tris-based buffer and KCl. This component regulates the pH of the solution which affects the activity and fidelity of the polymerase enzyme.

⊙ ***Taq* polymerase** – This is a thermostable enzyme isolated from the bacterium *Thermus aquaticus* that will perform the DNA synthesis/polymerization. Too little of this enzyme will yield too little product, and too much enzyme can result in the generation of artifacts that might cause your confirmation band to smear.

⊙ **dNTP mix** – This is a mix of deoxyribonucleotide triphosphates that the polymerase will use to amplify our target sequence. The mix includes all four dNTPs: adenine (A), guanine (G), cytosine (C), and thymine (T).

⊙ **MgCl$_2$** – This is a required cofactor for the *Taq* polymerase, and its concentration is key to successful amplification. If MgCl$_2$ is not supplied in adequate amounts, then the enzyme will be rendered ineffective or inactive. However, too much MgCl$_2$ will reduce the fidelity of the *Taq* polymerase.

⊙ **FWD primer** – This is a short strand of DNA that defines the region of the genome to be amplified. The primers we will be using, both forward (FWD) and reverse (REV), are 19 bases long and have a GC content of 42%. The FWD primer will bind/anneal to the antisense/reverse strand of the template DNA. The **sense DNA strand** is the one that is complimentary to the RNA that goes off to be used to make proteins, or, in our case, to be used as part of the 16S ribosomal subunit. Our FWD primer sequence is:

    5'-CGG TTA CCT TGT TAC GAC TT-3'

⊙ **REV primer** – This is the primer which will bind to the forward strand of the template DNA. The sequence for our REV primer is:

    5'-AGA GTT TGA TCC TGG CTC AG-3'

⊙ **BSA** – Bovine serum albumin. This is used for two purposes: 1) as a blocking agent that keeps the polymerase, primers, and dNTPs from binding to the tube walls or pipette tips, and 2) as an adjuvant that makes the DNA more accessible to the polymerase.

PROCEDURE

First, we make the **master mix.** ****Note**** *All the components should be **kept on ice** until they are needed, then placed immediately back in ice.* Add the following to a 0.2 ml PCR tube:

	VOLUME (µL)
Autoclaved diH$_2$O	15.55
MgCl$_2$	1.00
FWD Primer	1.00
REV Primer	1.00
dNTPs	2.50
PCR buffer	2.50
BSA	0.25
Taq polymerase	0.20
Final volume	24.00

Once all the components of the master mix have been added to the PCR tube, gently pipette up and down to homogenize the mix.

Once the master mix has been homogenized, add 1.00 µl of your extracted DNA to 24 µl of the MasterMix in a PCR tube and ensure adequate homogenization by gently pipetting the mixture up/down a few times. Ensure the PCR tube is tightly closed and there are no bubbles in your tube. Now you are ready to put your sample into the thermocycler. While in the thermocycler, the template DNA will be subjected to a series of temperature changes that facilitate the amplification of the target DNA sequence. Essentially, the artificial environment created in the thermocycler functions to mimic that of which would occur *in vivo*. However, here we are using heat to function as the DNA helicase and the DNA gyrase. See **Figure 1** below for a generalized schematic of a typical PCR cycle.

Thermocycler Settings

Denaturation: ~94°C for 5 min

Annealing: ~52 °C for 35 sec

Elongation: ~72 °C for 2 min

Number of cycles: ~35–40 cycles

Once the PCR process is complete, you will confirm that you have a PCR product by subjecting your sample to gel electrophoresis in a 0.8% agarose gel. This gel will be prepared by dissolving 400 mg agarose per 50 ml of sodium boric acid (SBA) buffer. You will load a stained portion of your product into one of the holes (wells) in the gel. On a piece of parafilm, mix 4 µl of your PCR product with 2 µl of SYBR Green. Load the entire 6 µl of your stained sample into an empty well in the gel — **MAKE NOTE OF WHAT WELL YOUR SAMPLE IS IN!** *Be very careful here.* Be sure that you load your sample directly into the well without puncturing the well. If you loaded your sample correctly, you will notice that the entire sample is at the bottom of the well. Electrophorese the gel in SBA buffer at 120 V for 20 min and inspect it with UV light to see if your lane has a glowing band of DNA. For our purposes, we are looking for a band around 1 Kbp in size.

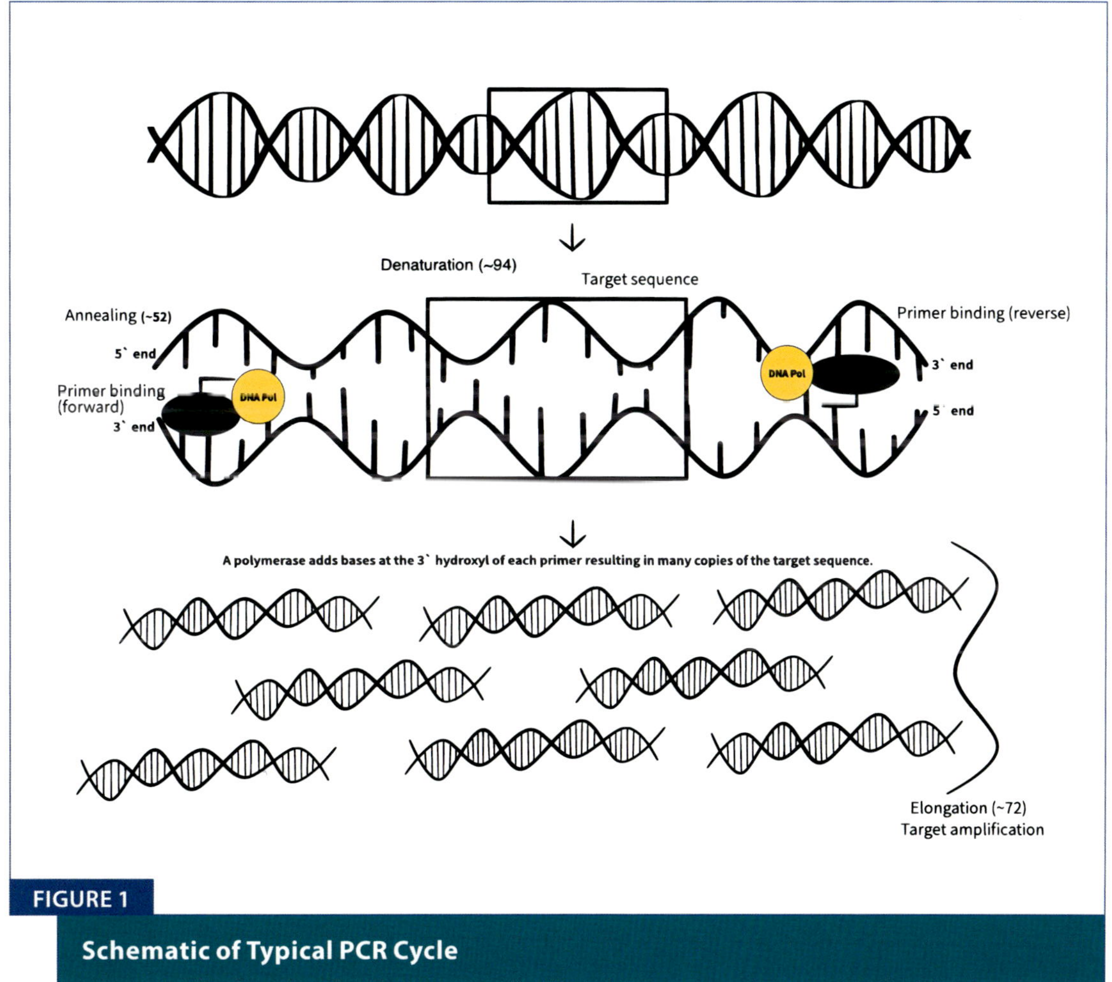

FIGURE 1

Schematic of Typical PCR Cycle

PART 3: RECEIPT & ANALYSIS OF SEQUENCING DATA

Roughly 15 µl of your PCR product will be sent to Eurofins Genomics in Louisville, KY, where your PCR product will be sequenced. Eurofins Genomics employs the Sanger Sequencing method, which utilizes and exploits the selective use of **dideoxynucleotides** (ddNTPs). These ddNTPs have been modified so that they are missing the 3'-OH group, which prevents the DNA polymerase from forming a phosphodiester bond. Because they lack that functional group, ddNTPs terminate DNA replication when they are incorporated into the newly synthesized DNA molecule. Thus, Sanger sequencing is often referred to as "Chain Termination Sequencing."

Your sample will be subjected to PCR again when it arrives at the sequencing company but instead of using only dNTPs, they will run four aliquots of your sample with each containing a mix of dNTPs and *one* type of ddNTP (A, T, C, or G). This will result in the synthesis of DNA chains of many different lengths, all terminating at different places according to when each ddNTP was incorporated. Visualization of DNA fragments and their ddNTPs is accomplished by using radioactively or fluorescently labelled ddNTPs and electrophoresing the PCR products (**Figure 2**).

Eurofins Genomics will send a digital file that contains your sequencing results, which will include the sequence of the amplified DNA (16S rRNA gene). To make use of this information, you need to compare your sequence to other bacterial species and determine if there are any sequences that are similar to your sequence. To perform this analysis, you can use a homology-based method to identify the genus of your bacterium. The tool we will use to perform this analysis is the **Basic Local Alignment Search Tool** (**BLAST**) curated by the National Center for Biotechnology Information (NCBI), a subsidiary of the National Institutes of Health (NIH).

The first version of BLAST was published in 1989. All versions use a set of statistical algorithms to compare DNA or protein **input (query) sequences** to a database of **subject sequences** and return measures of similarity between the query and a set of subject sequences. BLAST is fairly sophisticated and has several different applications, but we will use BLASTn, which compares a query DNA sequence (your sequence) to subject DNA sequences (curated sequences) in the NCBI GenBank database.

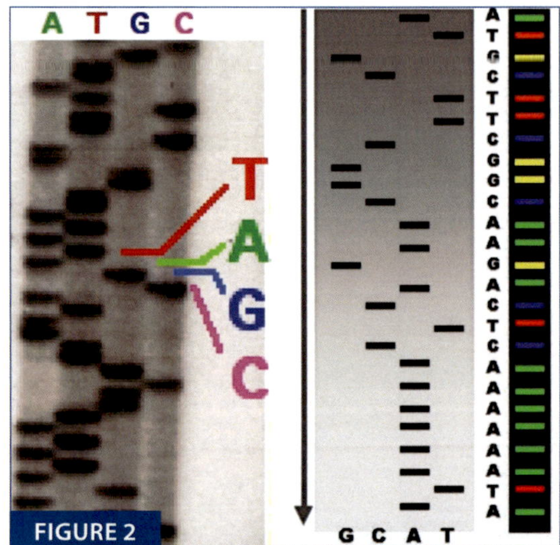

FIGURE 2

Two Examples of Sequencing Gels. On the left is a radiolabeled sequencing gel and on the right is a fluorescently labelled gel. In both, each column was created by one of the ddNTP PCR aliquots. Note that you can read the sequence by simply filling in each letter according to where there is a band. Left by John Schmidt [CC BY-SA 3.0]. Right by Abizar at English Wikipedia [CC BY-SA 3.0]

BLAST URL:

https://blast.ncbi.nlm.nih.gov/Blast.cgi?PROGRAM=blastn&PAGE_TYPE=BlastSearch&LINK_LOC=blasthome

GenBank was started in 1982 and as part of NIH, it is publicly-funded and publicly-available. Data are regularly exchanged between it, the **DNA Databank of Japan (DDBJ)**, and the **European Nucleotide Archive (ENA)**, so the results reflecting your query sequence will have the global research community behind it. The number of sequences in the database has doubled every ~18 months since its inception. GenBank Release 230, released in February 2019, contains 303,709,510,632 bases and 212,260,377 sequences. It is difficult to get current numbers on identified species, but in 2009 the database identified more than 300,000 organisms to the genus level or lower (species and strain).

You will use BLASTn to perform a nucleotide-to-nucleotide comparison between your sample sequences and sequences in GenBank. Essentially, the BLASTn algorithm will search for the top sequence-to-sequence matches between your query sequence and the sequences in the GenBank database. The basic procedure is to copy and paste your sequence into BLASTn as a query. To limit the BLASTn search time, you will then choose to compare your sequence to the 16S ribosomal RNA sequence database instead of a comparison between your sequence and all nucleotide sequences in the GenBank database. Once the comparison is made, BLASTn will indicate the length of your sequence and the top 100 bacterial species (hit table) most similar to your sequence. (See handout for further details.)

The results will contain several bits of information that you will use to identify your bacterium:

- **Similarity score** – How similar your sequence is to each of the species in the hit table.

- **Query coverage** – A report of what percentage of your query sequence overlapped the given subject sequence.

- **E-value** – An estimate of background noise. The lower the value, the more significant the match is. E-value describes how many hits one might "expect" to see when searching a database of a particular size. As your similarity score increases, E-value decreases exponentially. If your query was assigned an E-value of 3, that would mean you would expect to see 3 matches purely by chance.

- **Max identity** – The percent similarity of your query <-> subject over the length of the coverage area.

PART 4: GENOME ANNOTATION VISUALIZATION

Genome annotation refers to the identification of protein-coding regions within genomic DNA sequences and the prediction of protein functions based on the genetic code. In addition to predicted protein functions, these programs provide a glimpse into the genome that supplies other important information such as gene number, length of intergenic regions and their distance between genes, and many more critical pieces of information. There are an array of gene/genome annotation platforms that perform the annotation process using slightly different parameters; however, each platform will attempt to functionally characterize genes and gene products using homology-based references to predict gene and subsequent protein functionalities. Gene/genome annotation is a powerful tool that can supply instrumental insight into the identity of your organism, gene conservation between closely or distantly related organisms, and elucidate phenotypic expression.

To derive meaning from the annotation data, we must enlist the aid of a genome visualization program. Genome visualization programs such as **Artemis** facilitate a more comprehensive illustration and understanding of how genomic data are compiled. For example, Artemis illustrates where genes and intergenic regions arc located along the chromosome, which is essential for targeting regions of DNA in which we wish to amplify via PCR. In addition, Artemis enables you to perform keyword searches for genes of interest, which can expedite your research insofar as allowing you to quickly locate genes, determine if the gene is in the correct reading frame, identify important promoter regions/protein binding regions, as well as provide a predicted gene product for each gene. In short, genome visualization tools combine sequence and annotation data in a way that offers an interpretable means to rapidly expedite genetic and genomic studies.

Although the entire genome of your bacterial species will not be completely sequenced or annotated, it is essential to understand the link between the genetics undergirding the phenotypic expression of your specimen, i.e. the results you gained from performing various biochemical tests on your bacterial specimen. Therefore, you will download and analyze the complete genome sequence/annotation of the bacterial species that 1) shares the most 16S sequence homology with the 16S sequence data of your bacterial species, and 2) has a completely sequenced and annotated genome. The steps for downloading your corresponding genome are as follows:

1. Click on the top BLASTn hit associated with your 16S data.

2. Search for the top BLASTn hit in the NCBI genome database by entering the genus and species name of that organism.

3. Click the link associated with **Organism Overview** titled "Genome Assembly and Annotation Report."

4. Here, we want a complete genome, so we will look for a completely black circle under the level of completeness indicating that we've selected a genome that is completely sequenced and completely annotated. Once we've found this, we will click on the associated 'chromosome' link under the Replicons column, e.g. NC_002516.2.

5. Next, you will see a GenBank link (in blue) with a drop-down arrow. Click on this and select the "GenBank (full)" option.

6. Then pan over to the "Send to" link and ensure that the "complete record" option is selected and then select the "file" option under "Choose Destination" and click the Create File link.

7. You should now have the complete genome sequence and annotation of that bacterial species that is in GenBank format, e.g., sequence.gb

Now that you have the annotated sequence downloaded, you need a program that will interpret the genomic data and will allow you to visualize how the genes and gene products are represented within the genome (**see additional handout for downloading Artemis Genome Browser**).

We will discuss the functionalities of Artemis during class in greater detail and will collectively engage in a problem set that will

1. promote familiarity with the program, and

2. strengthen our knowledgebase on the following microbial genetic/genomic principles: DNA replication, transcription, translation, gene conservation, and protein conservation.

REFERENCES

Promega (2019). PCR Amplification – Protocols and background information about PCR and RT-PCR. Online resource: https://www.promega.com/resources/product-guides-and-selectors/protocols-and-applications-guide/pcr-amplification/ (Accessed July 10, 2019).

Plant & Soil Sciences eLibrary Pro (2019). Polyerase chain reaction (PCR). University of Nebraska Lincoln. Online resource: http://passel.unl.edu/pages/informationmodule.php?idinformationmodule=968252315&topicorder=3&maxto=11, (Accessed July 10, 2019).

Woese, C.R. & Fox, G.E. (1977). Phylogenetic structure of the prokaryotic domain: The primary kingdoms. PNAS, Nov 1, 1977 74:11 pp 5088-5090.

NIH (2019). GenBank & WGS Statistics. https://www.ncbi.nlm.nih.gov/genbank/statistics/

Benson DA, Karsch-Mizrachi I, Lipman DJ, Ostell J, and Sayers EW (2010). GenBank. Nucleic Acids Research 2010 Jan; 38(Database issue): D46–D51. doi: 10.1093/nar/gkp1024

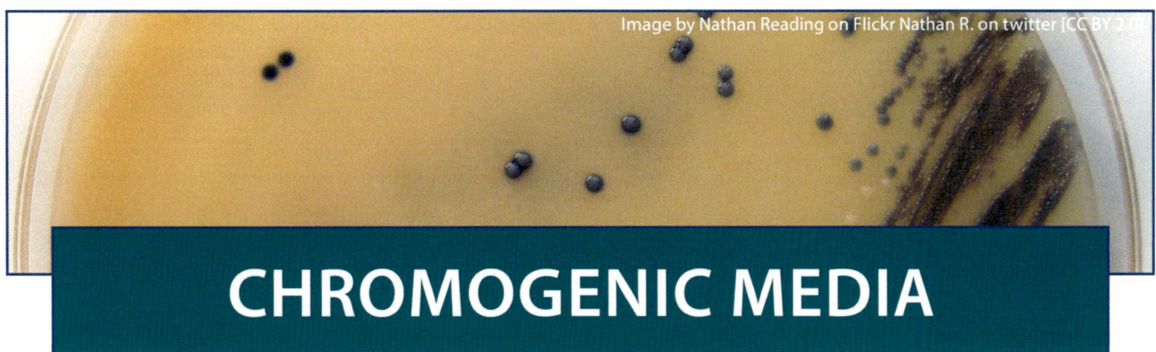

CHROMOGENIC MEDIA

The traditional approach to identifying bacteria, especially in clinical contexts, has, since the late-1800's, included the appearance of colonies on solid media and use of various microbiological media designed to detect biochemical differences. **Serological methods,** which involve the use of antibodies to detect certain pathogen-specific **epitopes,** have also commonly been used since the early-mid 20th century, though they typically come into play only once the initial biochemical screening has been completed. This "screening" is time and resource intensive. These facts have led, over the last 30 years, to the development of new, rapid screening tests. These tests are done using chromogenic media.

Chromogenic media are microbiological growth media that have been designed to target pathogens with a high-specificity. These media accomplish this through the use of specially-designed enzyme substrates (**chromogens**) that, when hydrolyzed, are reduced in solubility and precipitate, imparting a color to the bacterial colonies. This method relies on the assumption that your given pathogen expresses a given enzyme, but as you know, bacteria are highly diverse. It is therefore possible that some strains of any particular pathogen might lack the target enzyme, and so a second type of chromogenic media is used: one that targets the same pathogen but via a different enzyme. Even so, some of these media, like the one we will be using today, have sensitivities of up to 95.5% for certain species.

It is also common for chromogenic media to contain selective elements as well. These compounds, when present, are usually tailored to inhibit commensal bacterial species. Commensal species that are not inhibited would ideally grow to have colorless colonies. This combination of differential and selective properties has the effect of dramatically reducing both the time and materials needed to screen a clinical specimen for one or more pathogens. They can also serve to decrease the **limit of detection** because the colored colonies are much easier to pick out.

In this lab exercise, as part of your ongoing effort to positively identify your unknown organism, you will apply your organism to CHROMagar™ Orientation. This medium has been designed to allow for the identification of following species via the following visual cues:

TABLE 1

INTENDED SPECIES	RESULT
Proteus vulgaris	Clear colonies with a brown halo in surrounding medium
Pseudomonas aeruginosa	Transparent colonies light yellow in medium
Staphylococcus aureus	Yellow colonies, less translucent
Escherichia coli	Purple colonies which bleed into the medium
Klebsiella pneumoniae	Dark blue colonies which bleed
Enterococcus faecalis	Lighter turquoise blue that bleeds
UNINTENDED SPECIES	**RESULT**
Salmonella typhimurium	Translucent mucoid colorless colonies like *P. aeruginosa* but without the yellow
Alcaligenes faecalis	Translucent colorless colonies like *S. typhimurium* but less mucoid
Klebsiella aerogenes	Translucent gray/blue colonies
Bacillus cereus	Waxy white colonies
Bacillus thuringiensis	Waxy light turquoise colonies
Micrococcus luteus	Opaque white colonies
Pseudomonas fluorescens	Transparent colonies with a yellow in the medium like *P. aeruginosa* but more of it

PROCEDURE

1. Obtain your maintenance culture and one plate of CHROMagar™ Orientation medium.

2. Transfer a colony to the specialized medium.

3. Incubate at 37° C for 24 hrs.

4. After incubation, observe any colonies growing in the plate and compare your results to Table 1.

REFERENCES

Perry JD, Freydière, AM (2007). The application of chromogenic media in clinical microbiology. *Journal of Applied Microbiology*. December 2007 103:6, pp 2046–2055. https://doi.org/10.1111/j.1365-2672.2007.03442.x.

EVALUATING DISINFECTANTS & ANTISEPTICS

The effectiveness of disinfectants and antiseptics does not rely on active metabolism, so they can be used to control viruses, bacteria that produce endospores, regular vegetative bacterial and fungal cells, as well as fungal spores. These compounds are fairly reactive, and they are not selective in their activities so they are limited to use on surfaces (disinfectants), including the outside of organisms (antiseptics). You can find in Table 1 a list of some common disinfectants and antiseptics, their modes of action, and usages.

- **Disinfectants** – Are more reactive than antiseptics so their use is generally restricted to non-living surfaces where tissue damage is not a danger.

- **Antiseptics** – Are somewhat less reactive than disinfectants and can be used on the outside of living things. Some antiseptics, like mouthwash, can be used in natural body openings, which is not what many people would think of when they hear the words "external use only."

There are some factors that you need to take into account when you set out to use disinfectants and antiseptics. The manufacturer of these products will often provide some insight into these influential factors the form of usage instructions. In general, you need to consider the following:

- **Concentration** – The concentration of an antimicrobial chemical has a significant impact upon its efficacy. In many cases, more chemical means quicker or more effective lethality, but that is less true for things like isopropyl alcohol which are more effective if they are diluted with a little water. More chemical might also mean tissue or surface damage, so follow the usage instructions and don't be a cowboy (or girl).

- **Duration of exposure** – Chemical reactions are not instantaneous; they actually take some time to proceed. Increasing the duration of the exposure increases the number of chemical reactions that can and do occur.

- **Type of microbe you intend to destroy** – Not every microbe is the same. Trying to disinfect a surface of an enveloped virus like HIV may actually require less work than disinfecting a surface of a spore-forming bacterium or a yeast. Spores (bacterial or fungal), **encapsulated bacteria, acid-fast bacteria,** and naked viruses are more hardy.

- **Environmental conditions**

  - **Temperature** – Temperature controls how fast chemical reactions occur. Raise the temperature and the rate of chemical reactions goes up, lower the temperature and the rate of reaction goes down.

  - **pH** – Some disinfectants are more effective at extreme pHs, and some require an environment fairly close to neutral to prevent the compound from ionizing into a less active form.

  - **Type of material you are treating** – Non-porous surfaces usually require less concentration and less contact time to achieve the same level of **asepsis** as a porous surface.

In this laboratory, you will use a version of the disk diffusion method for evaluating which disinfectants and antiseptics are effective at inhibiting the growth of some bacterial species. Beware during your interpretations that you do not fall into thinking that a larger clear area means more efficacy as it does in the Kirby–Bauer test. We are not controlling for different diffusion rates in this disk diffusion method. Hence, some chemicals may diffuse through the medium more quickly than others which would lead to a larger area of inhibited growth. We are also not adequately testing for the amount of the various compounds contained in each filter disk. What we *can* determine is a bit narrower: We can determine whether any of the compounds are effective. Once we know which are effective, we could do a more targeted test that controls for diffusion rates, such as the use-dilution test.

PROCEDURE

1. Obtain the bacterial cultures, one TSA plate for each, and a sterile swab for each.

2. Label each TSA plate with the organism you will inoculate into it.

3. Make a spread plate with 100 µl of bacterial culture. Be sure to get as even a coverage as you can.

4. You will then place filter papers that have been wetted with each of various antimicrobial compounds onto the plate.

   a. Label each plate with the compounds you will be testing.

   b. Wet a sterile filter paper using one of the compounds.

c. Blot the wet filter paper between layers of a paper towel to squeeze out excess liquid. If you leave too much liquid in the paper, a larger amount of compound will diffuse out into the medium and can drown out the results associated with adjacent filter papers.

d. Using forceps, place a blotted filter paper onto your prepared spread plate in its properly-labelled spot and press firmly down on it with the forceps to ensure it adheres to the medium surface.

e. Once all the papers are placed, parafilm the plate and leave it face-up for several minutes to ensure the filter papers all finish adhering to the surface. If you fail to wait, your filter papers may fall off the medium surface when they are placed into the incubator in the standard inverted position.

5. Incubate the plates at 37° C for 24 hrs.

6. After incubation, observe the plates and note any filter papers with a clear zone around them. These are compounds that were effective and warrant a more targeted test.

Common Disinfectants and Antiseptics

TABLE 1

AGENT	MODE OF ACTION	USE
Phenolic Compounds		
Phenol	Protein denaturation; Precipitation of proteins; Membrane disruption	0.5%–1% solutions antiseptic and mild local anesthetic; 5% solution is disinfectant
Cresols	Similar to phenol	Active ingredient in Lysol; 2%–5% Lysol is disinfectant
Hexachlorophene	Similar to phenol	Topical antiseptic effective against gram + organisms; Beware of neurotoxic effects
Hexylresorcinol	Similar to phenol	Mild local anaesthetic; Topical treatment of small skin infections, anti-aging creams; Beware of estrogen-like effects
Thymol	Similar to but less effective than phenol	Antifungal; Effective against hookworm; Diluted solutions as mouthwash/gargle
Alcohols		
Alcohols	Lipid solvent; Denaturation/coagulation of proteins; Wetting agent in tinctures; Effect increases with molecular weight	Skin antiseptic; Ethyl 50–70%; Isopropyl 60–70%
Halogens		
Chlorine Compounds	Combination with protein; Forms hypochlorous acid in water; Oxidizer; Allosteric enzyme inhibitor	Sanitizing water; Sanitation of restaurant and dairy utensils; Chloramine 0.1–2% for wound irrigation and dressings
Iodine Compounds	Not fully understood; May precipitate protein; Surface-active agent	Tincture for skin antiseptic; Treatment of goiter; Effective against spores, fungi, and viruses.

Table 1 continued: Common Disinfectants and Antiseptics

AGENT	MODE OF ACTION	USE
Heavy Metals		
Inorganic	Mercuric ion precipitates protein; Allosteric enzyme inhibitor	Beware of irritation and systemic toxicity; Ineffective against spores; Laboratory disinfectant
Organic	Similar to inorganic but more useful	Skin antiseptic; Less toxic, less irritating; Ineffective against spores
Silver nitrate	Precipitation of proteins; Metabolic blocker	Antiseptic of mucous membranes
Surface-Active Agents		
Emulsifiers, soaps, detergents	Lower surface tension of water; Membrane disruption; Alteration of cell permeability; Alter pH of environment	Weak activity against fungi, acid-fast organisms, spores, and viruses
Cationic Agents (quaternary ammonia)	Similar to other surface-active agents; Activity reduced by soaps	Bactericidal, fungicidal; Ineffective against spores and viruses; Asepsis of unbroken skin; Disinfectant for operating-room equipment, dairies, and restaurants
Anionic agents (sodium tetradecyl sulfate)	Similar to other surface-active agents; Activity optimal in acidic conditions	Sclerosing agent in treatment of varicose veins and internal hemorrhoids
Alkylating Agents		
Formaldehyde (gas or liquid)	Adds alkyl groups to nucleic acids and proteins; Crosslinks connective tissues	Active against bacteria, fungi, and viruses; Disinfection of rooms; In alcohol for instrument disinfection; Preserving biological specimens; Deactivation of viruses for vaccine development
Ethylene Oxide	Same as other alkylating agents	Sterilization of materials that would be damaged by heat
B-Propiolactons	Same as other alkylating agents	Sterilization of tissues before grafting; Deactivation of hepatitis viruses; Disinfection of rooms
Crystal Violet	Bonds to nucleic acids and peptidoglycan; Effective against gram+ organisms	Skin antiseptic; Treatment of thrush in infants; Laboratory isolation of gram− bacteria

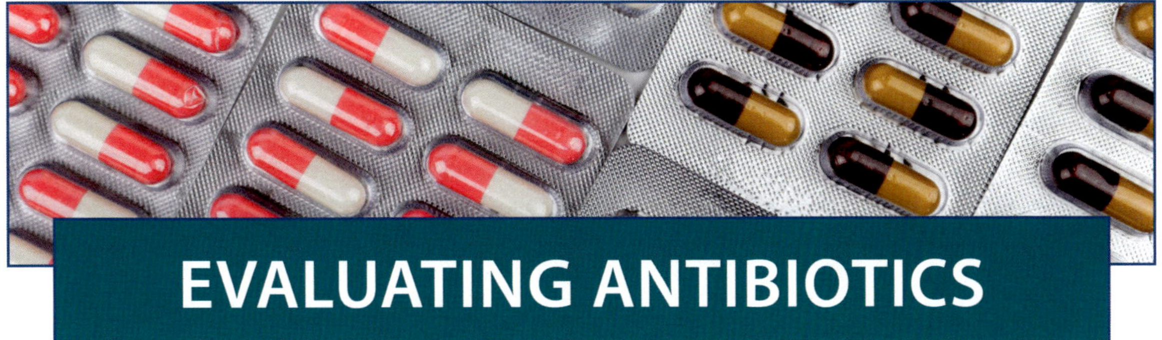

EVALUATING ANTIBIOTICS

Antibiotics are in a very different class than disinfectants and antiseptics, though they all fall under the broad category of **antimicrobial agents. Antibiotics** are antimicrobial drugs that exhibit **selective toxicity** wherein they are detrimental to a parasite and not to the parasite's host. **Disinfectants** and **antiseptics** are non-selective chemicals that have broad reactivity to a variety of organic compounds.

Antimicrobial compounds work in one or more of these very general ways:

- Interference with cell walls.
- Interference with cell membranes.
- Alteration of the colloidal state of cytoplasm.
- Interference with cellular enzymes.
- Interference with the structure or function of DNA.
- Interference with protein synthesis.

ANTIBIOTICS

Antibiotics were originally derived from living microorganisms, mainly **eubacteria, actinomycetes,** and fungi. These organisms possess genes that code for various antimicrobial compounds, and they are capable of expressing those compounds as weapons against other microbes. Many of these microbial warfare compounds exist, but only a small number of them are sufficiently selective in their toxicity so as to allow their use in humans and animals. Unlike disinfectants and antiseptics, antibiotic toxicity is dependent upon there being active cellular metabolism in the targeted organism. A list of common antibiotics, their **modes of action,** and potential side effects is given in Table 1.

The **antibiotic resistance** exhibited by some microbes is a topic of great interest and has been since shortly after the widespread use of antibiotic compounds. There are a number of different ways that organisms may exhibit resistance to antibiotics, including the following:

- ⊙ Efflux pumps that pump out antibiotic compounds before they can harm the cell.

- ⊙ Expression of enzymes that destroy the antibiotic.

- ⊙ Alteration of metabolic pathways to exclude susceptible steps.

- ⊙ Change in the conformation of the antibiotic target, such as an enzyme or ribosome.

Additionally, there are two classes of antibiotic-like drugs:

- ⊙ **Synthetic:** they have no origin in microbes and were constructed wholly in laboratories. These often work as **competitive inhibitors** wherein they slow enzymatic reactions.

- ⊙ **Semi-synthetic:** the organism that makes the compound must first be fed a synthetic precursor compound that is not found in nature. The microbe modifies the synthetic precursor and produces the final, active product.

In this laboratory, you will use a modified Kirby-Bauer antibiotic sensitivity test to determine how some bacteria respond to six antibiotics. This method controls for the amount of antibiotic (listed on each Sensi-Disc) and for speed of diffusion through the medium. We can therefore use the diameters of any resulting clear zones of inhibited growth to make quantitative determinations about the efficacy of each drug.

PROCEDURE

1. Obtain the organisms in question and a TSA plate for each.

2. Prepare a spread plate with 100 µl of bacterial culture. Be sure to get as even a coverage as you can.

3. Spray a paper towel with a bit of Lysol to make a Lysol wipe.

4. Also obtain a Sensi-Disc Dispenser and pressing down firmly *once*, dispense the six antibiotic pads. If you hesitate or short-stroke the dispenser, you can cause it to jam.

5. Wipe the plate-side of the dispenser with your Lysol wipe before passing it to the next person.

6. Once the Sensi-Discs have been placed on the medium surface, leave the plates face-up for several minutes before parafilming them. If you skip this step, the discs may not fully adhere to the medium surface and fall off when the plates are incubated in the standard inverted position.

7. Incubate plates at 37° C for 24 hrs.

8. After incubation, observe your plates, measuring the diameter of any clear zones in mm and compare your data to those in Table 2 to determine whether the organism is susceptible, intermediate, or resistant to each antibiotic.

Common Antibiotics

TABLE 1

AGENT	USE	MODE OF ACTION	SIDE EFFECTS	SPECTRUM
Ampicillin	UTI, URTI, menigitis, salmonellosis	Penicillin type, wider effectiveness in gram− cells	Rash, nausea, diarrhea, hairy tongue, colitis; Side effects more likely in those with asthma and allergies, especially penicillin allergy.	Broad
Azithromycin	Strep throat, gonorrhea, chlamydia, Legionnaire's disease	Protein synthesis inhibition; binds 50S ribosomal subunit, inhibiting translation	Diarrhea, nausea, abdominal pain, pseudomembranous colitis.	Broad
Chloramphenicol	Eye ointment, plague, cholera, typhoid fever	Inhibits protein synthesis; binds to 50S ribosomal subunit and interrupts protein chain elongation.	Headache, depression, confusion. Aplastic anemia, bone marrow suppression, hypersensitivity reaction.	Broad
Ciprofloxacin	UTI, RTI, typhoid fever, anthrax, chancroid	DNA synthesis inhibition; interferes with DNA gyrase and some topoisomerases.	Nausea, diarrhea, abnormal liver function, vomiting, rash, tendinitis.	Broad, but greater against gram− and less against gram+
Clindamycin	Anaerobic infections, including dental infections	Protein synthesis inhibition; binds 50S ribosomal subunit, inhibiting translation.	Diarrhea, pseudomembranous colitis, nausea, vomiting, abdominal pain/cramps, rash. Metallic taste, vaginal fungal infection.	Aerobic gram+, anaerobic gram−
Erythromycin	Infections of skin and respiratory tract	Not fully understood; binds 50S ribosomal subunit, may prevent the binding of tRNAs.	Diarrhea, pseudomembranous colitis, nausea, vomiting, abdominal pain/cramps, rash. Arrhythmia, anaphalaxis.	Aerobic gram+ and gram−
Gentamicin	Serious infections only	Inhibits protein synthesis; irriversibly binds to 30S ribosomal subunit.	Low blood counts, allergy, neuromuscular anomolies, nerve damage, nephrotoxicity, ototoxicity.	Gram− organisms and Staphylococcus
Methicillin	No longer prescribed	Penicillin type	Formerly used for infections caused by penicillin-G-resistance. No longer manufactured because other, more stable options are available such as cefoxitin.	Gram+ organisms
Penicillin G	Syphilis, strep throat, gas gangrene, leptospirosis and many others	Inhibits bacterial cell wall synthesis by irreversibly binding to DD-tanspeptideases, stopping the final peptidoglycan crosslinking.	Diarrhea, nausea, and hypersensitivity responses including fever, joint pains, urticaria, rashes, angioedema, anaphylaxis	Gram+ plus some gram− outliers in Neisseria and Leptospira
Rifampin	MRSA, Tuberculosis, leprosy, Legionnaire's disease, Lyme disease	Inhibition of RNA synthesis by binding to RNA polymerase and blocking RNA transcript elongation	Hepatotoxicity, breathlessness, flushing, rash, diarrhea, nausea,	Broad

Table 1 continued: Common Antibiotics

AGENT	USE	MODE OF ACTION	SIDE EFFECTS	SPECTRUM
Streptomycin	Plague, tuberculosis, endocarditis	Inhibition of protein synthesis by binding to the 30S ribosomal subunit causing codon misreads	Nephrotoxicity, ototoxicity (specifically to cranial nerve VIII), not to be used during pregnancy but fine while breast feeding.	Broad
Sulfonamides	UTI, bronchitis, bacterial meningitis, ear infections	Slows folate synthesis by competitively inhibiting dihydropteroate synthase	Allergic hypersensitivity	Aerobes
Tetracycline	Acne, cholera, plague, malaria, syphilis	Inhibition of protein synthesis by binding to the 30S ribosomal subunit, blocking peptide elongation	Diarrhea, vomiting, rash, loss of apetite, poor tooth development in those younger than 8 yrs old, nephrotoxicity, increased likelihood of sunburn.	Broad
Vancomycin	Last resort for skin, blood, and endocardial infections	Inhibits bacterial cell wall synthesis by irreversibly binding to un-crosslinked D-alanyl-D-alanine, stopping the final peptidoglycan crosslinking.	Allergic hypersensitivity, nephrotoxicity, ototoxicity, low blood pressure, bone marrow suppression, red man syndrome.	Gram+

Results Interpretation Table

TABLE 2

AGENT	DISK CONTENT	ZONE DIAMETER TO THE NEAREST WHOLE mm		
		SUSCEPTIBLE	INTERMEDIATE	RESISTANT
Ampicillin				
Gram−	10 µg	≥ 17	14–16	≤ 13
Gram+	10 µg	≥ 17	−	≤ 16
Azithromycin				
Gram−	15 µg	≥ 13	−	≤ 12
Gram+	15 µg	≥ 18	14–17	≤ 13
Carbenicillin				
Chloramphenicol	30 µg	≥ 18	13–17	≤ 12
Ciprofloxacin				
Gram−	5 µg	≥ 31	21–30	≤ 20
Gram+	5 µg	≥ 21	16–20	≤ 15
Clindamycin	2 µg	≥ 21	15–20	≤ 14
Erthyromycin	15 µg	≥ 23	14–22	≤ 13
Gentamicin				
Enterobacteriaceae	10 µg	≥ 15	13–14	≤ 12
Staphylococcus	10 µg	≥ 15	13–14	≤ 12
Methicillin (cefoxitin)				
Enterobacteriaceae	30 µg	≥ 18	15–17	≤ 14
Staphylococcus	30 µg	≥ 25	—	≤ 24
Penicillin G	10 µg	≥ 29	—	≤ 28
Rifampin	5 µg	≥ 20	17–19	≤ 16
Streptomycin	10 µg	≥ 15	12–14	≤ 11
Sulfonamides	250 or 300 µg	≥ 17	13–16	≤ 12
Tetracycline				
Gram−	30 µg	≥ 15	12–14	≤ 11
Gram+	30 µg	≥ 19	15–18	≤ 14
Trimethoprim	5 µg	≥ 16	11–15	≤ 10
Vancomycin	30 µg	≥ 17	15–16	≤ 14

Data Source: Clinical and Laboratory Standards Institute. *Performance Standards Antimicrobial Susceptibility Testing, 28th edition (2018).*

Antibiotics Used

TABLE 3		
ANTIBIOTIC	**MASS ON DISK**	**LABEL**
Streptomycin	10 µg	S 10
Sulfadiazine	0.25 mg	SD .25
Penicillin	10 IU	P 10
Gentamicin	10 µg	GM 10
Vancomycin	30 µg	VA 30
Chloramphenicol	30 µg	C 30

MICROBIOLOGY OF PERSONAL CARE PRODUCTS

Macroorganisms, including humans, carry with them a panoply of **natural microflora.** You are carrying this collection of microbes on your skin, in your mouth, your upper respiratory tract, your urogenital tract, and your colon. What counts as "natural" varies depending on body location, between individuals, between geographical locales, and through time. This means that any object that comes into prolonged or significant contact with any part of your body or natural body openings is likely to carry away with it a sample of your microflora.

These organisms are living in some type of **symbiosis** with you. These symbioses may be one of three types: **commensal, mutualistic,** or **parasitic.** In commensal symbiosis, one party derives some benefit from the relationship, and the other derives nothing either positive or negative. In mutualism, both parties benefit from their association. In parasitism, one party benefits while the other is harmed

In the case of your microflora, the vast majority of them are commensal or mutualistic. Some common natural microflora bacterial species come from the genera *Staphylococcus* (skin), *Streptococcus* (urogenital/pharyngeal), *Neisseria* (pharyngeal), *Enterobacter* (colon), and *Proteus* (colon). There is a chance, however, that you may be carrying parasitic (pathogenic) organisms such as Group A β-hemolytic *Streptococcus, Serratia marcescens,* or Methicillin-resistant *Staphylococcus aureus* (MRSA).

In this laboratory exercise, you will inoculate a solid medium with one of your own personal care items such as a razor, deodorant stick, toothbrush, loofa, make-up brush, or something else that has prolonged contact with your body. A wristwatch or glasses will work in a pinch. We will be incubating these organisms at 37° C, average human body temperature. Under these conditions it is possible that some samples may come back positive for one or more organisms that are potentially pathogenic.

You must be very careful to observe every facet of aseptic technique when working with the plates after incubation. The colonies on the plates are comprised of many millions of cells, which is much more than what you would normally be exposed to. Thus, an organism that is not commonly pathogenic may cause an infection just by virtue of the extremely high concentration of cells which can overwhelm your immune response.

The medium you will use in this exercise is mannitol salt agar (MSA), which is a selective and differential medium. MSA is selective because it contains 7.5% NaCl which will inhibit the growth of most organisms except for halophiles such as *Staphylococcus*. It is differential because it contains the carbohydrate mannitol and phenol red, a pH indicator that will turn from red to yellow in the presence of acid. A bacterium originating from human skin, which is also halophilic and able to ferment mannitol, is likely to be *S. aureus*. Therefore, any colonies that turn the medium yellow are presumed to be *S. aureus* which is a potential human pathogen. Don't be alarmed if your sample turns up *S. aureus;* up to 30% of humans are carriers.

PROCEDURE

1. Obtain an MSA plate.
2. Inoculate the plate using your personal care item.
    a. If a razor or make-up brush, apply directly to medium surface.
    b. If powder makeup, tap brush on the edge of the plate to knock some dust onto the medium surface.
    c. If something like a loofa, apply directly to surface or use a swab to sample near the center.
    d. If something like glasses or a watch, use a swab to sample an appropriate surface.
3. Incubate at 37° C for 24 hrs.
4. Observe after incubation and determine whether any organisms grew and whether they are likely *S. aureus*.

EFFECTS OF pH ON GROWTH

Temperature is a factor that can affect the growth of microbes, but it isn't the only one. Microbial growth can be affected by pH as well but for different reasons.

pH is the quantitative measure of the acidity or alkalinity of a solution. It is theoretically equivalent to the following:

$$pH = -\log[H^+]$$

Where:

[ ] = concentration

H^+ = hydrogen ion

And:

pH < 7.0 = acidic

pH = 7.0 = neutral

pH > 7.0 = alkaline

Every microorganism has a pH range over which they are able to survive and reproduce, and these typically encompass 2–3 pH units. These organisms also have an optimal pH at which they will grow at their maximal rate. Natural environments are most commonly within the range of pH 3–9, so it shouldn't be surprising that most microorganisms that we have been able to culture have pH optima that lie within this range. As in our investigation of temperature's effects on microbial growth, we can classify microorganisms into three basic categories (Figure 1):

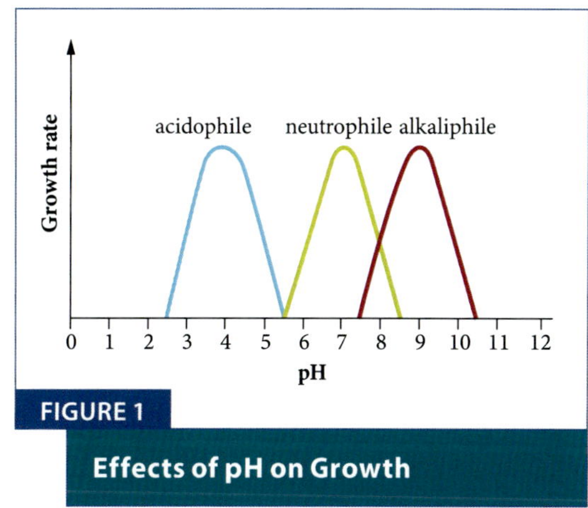

FIGURE 1

Effects of pH on Growth

- **Acidophiles** – Microorganisms whose pH optimum is below pH 5.5. For these organisms, a high concentration of H^+ is required for the cytoplasmic membrane to remain stable.

- **Neutrophiles** – Microorganisms whose optimum is between pH 5.5–7.9. These are the organisms that could live and grow in/on the human body.

- **Alkalophiles** – Microorganisms whose optimum is pH 8 and up. These organisms are interesting because some of them use Na^+ instead of H^+ to drive materials' transport and motility.

In this laboratory exercise, you will be investigating the pH ranges and optima of some bacterial species. You will do this using TSB to which pH **buffers** have been added to raise or lower the pH. Of course, it is worth noting that some of the integral parts of the medium can act as pH-stabilizing buffers such as amino acids, peptones, and proteins. We have added additional chemical buffers to drive the pH of the broth up or down and not merely to make the pH stable.

PROCEDURE

1. Obtain one tube of each TSB pH for each organism you are investigating. The pHs will be 3.0, 5.0, 7.0, and 9.0.

2. Inoculate each tube with an equal amount of inoculum.

3. Incubate the tubes at 37° C for 24 hrs.

4. After incubation, vortex one of your pH tubes and fill a cuvette with a subsample. Measure the optical density of the subsample using one the spectrophotometers. Now do the same for each of your pH tubes in turn.

    Be sure to prepare a blank for each pH broth. Their colors aren't equivalent, and that will yield misleading data if the spec is not properly zeroed.

EFFECTS OF TEMPERATURE ON GROWTH

Microbial growth is defined as an increase in the number of microbial cells over time, and this growth is highly dependent upon environmental temperature. This dependence is founded on the enzyme compliment that a particular microbe has. There are enzymes that work efficiently in temperatures near the freezing point of water, there are enzymes that work well at room temperature, and there are enzymes that can function above the boiling point of water. Thus any talk of "high" or "low" temperature must be relative because whatever temperature is high or low to one organism may be ideal for another and fatal to yet another.

Most **enzymes** are a class of proteins, and like all proteins, they have a 3D structure that is key to their functioning. If that 3D structure is altered, say through an increase in temperature that causes a shuffle in the secondary or tertiary bonding, then the enzyme will no longer function as it previously did. This change in 3D structure is called **denaturation,** and it can be temporary or permanent depending on the severity and duration of the temperature alteration. Permanent denaturation of cellular enzymes results in cell death.

Conversely, cold temperatures do not denature enzymes but rather slow the rate of their chemical reactions. As reaction times are slowed, the interactions between enzymes and substrates are also slowed. The result is a slower-than-optimal reaction rate and a correspondingly slower growth rate. If temperatures fall far enough, enzyme reaction rates can fall to zero, and microbial metabolism stops. This doesn't mean that the microbe is dead, however. If the cell retains its cell wall and membrane integrity, it is perfectly capable of resuming metabolism if the temperature rises.

Bacteria are not the only prokaryotes in the world, but they are the main focus of this laboratory. In bacteria, it is most common to find enzymes that will function somewhere within the range of –5° C to 80° C. It is common for a particular bacterium to have a growth temperature range of ~30° C though there are many that have narrower or wider ranges than this. These functional ranges are defined by **cardinal temperature points** (Figure 1):

- **Minimum growth temperature** – This is the lowest temperature that will allow some measurable amount of microbial growth. Below this temperature key cellular enzymes are inactive so no detectable growth occurs. This doesn't mean the microbes are dead. Many microbes are capable of producing chemicals that function like anti-freeze to keep ice crystals from forming in their cytoplasm. Others can create endospores that have a very low concentration of free water and are therefore impervious to freeze-killing.

- **Optimum growth temperature** – This is the temperature at which optimal growth occurs. Optimal growth doesn't mean that all the enzymes in the cell are working at their maximal rate.

- **Maximum growth temperature** – This is the highest temperature at which the microbial population will continue to grow. Above this temperature, enzymes are permanently denatured at such a rate as to make growth impossible, and the cells die.

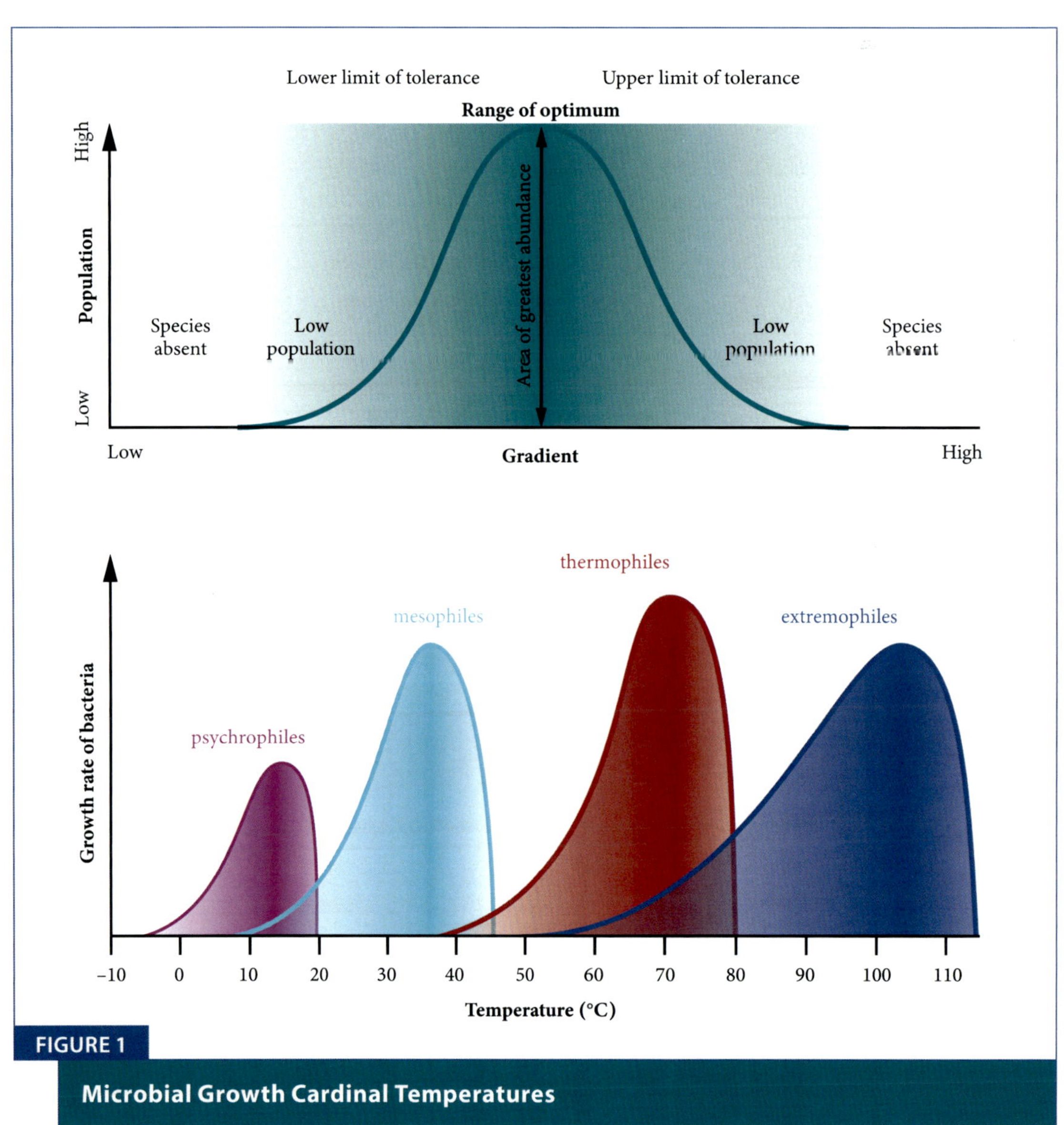

FIGURE 1

Microbial Growth Cardinal Temperatures

These temperature values can be used to define classes of bacteria. The three basic classifications are as follows:

- **Psychrophiles** – Microbial species that will thrive at ~10° C and generally will grow within a temperature of –5–20° C.

- **Mesophiles** – Microbial species that will grow at human body temperature (37° C) and not above 45° C. Their range is within 20–45° C. They can be split into two subgroups:

  - Those whose optimum lies within 20–30° C, which are usually plant **saprophytes.**

  - Those whose optimum lies within 35–40° C, which are generally organisms that prefer to grow in/on the bodies of **homeothermic** hosts like mammals.

- **Thermophiles** – Microbial species that will grow at 35° C and above. These can be split into two subgroubs:

  - **Facultative thermophiles** – Microbes that will grow at 37° C and with an optimum within 45–60° C.

  - **Obligate thermophiles** – Microbes that will grow only at temperatures in excess of 50° C and whose optimum is above 60° C.

There is a fourth classification category that can be useful: **Extremophiles.** These are microbial species that live and grow on the extreme ends of the environmental conditions found on earth. With regard to temperatures, there are cold extremophiles that will grow at temperatures within the range of –20 to –17° C, and there are hot extremophiles that will grow at temperatures within the range of 100° C to 121° C.

This week, you will be investigating the cardinal temperatures for various bacterial species.

PROCEDURE

1. Obtain cultures of the bacteria and enough tubes of TSB to incubate a sample of each species at the following temperatures: 4° C, 23° C, 37° C, and 60° C.

2. Mark your tubes with the species and incubation temperature.

3. Inoculate the TSB with each of the bacterial cultures, being careful to use the same amount of inoculum in each tube.

   - Be sure to vortex the stock cultures before you use them. This homogenizes the cell density to the greatest possible degree.

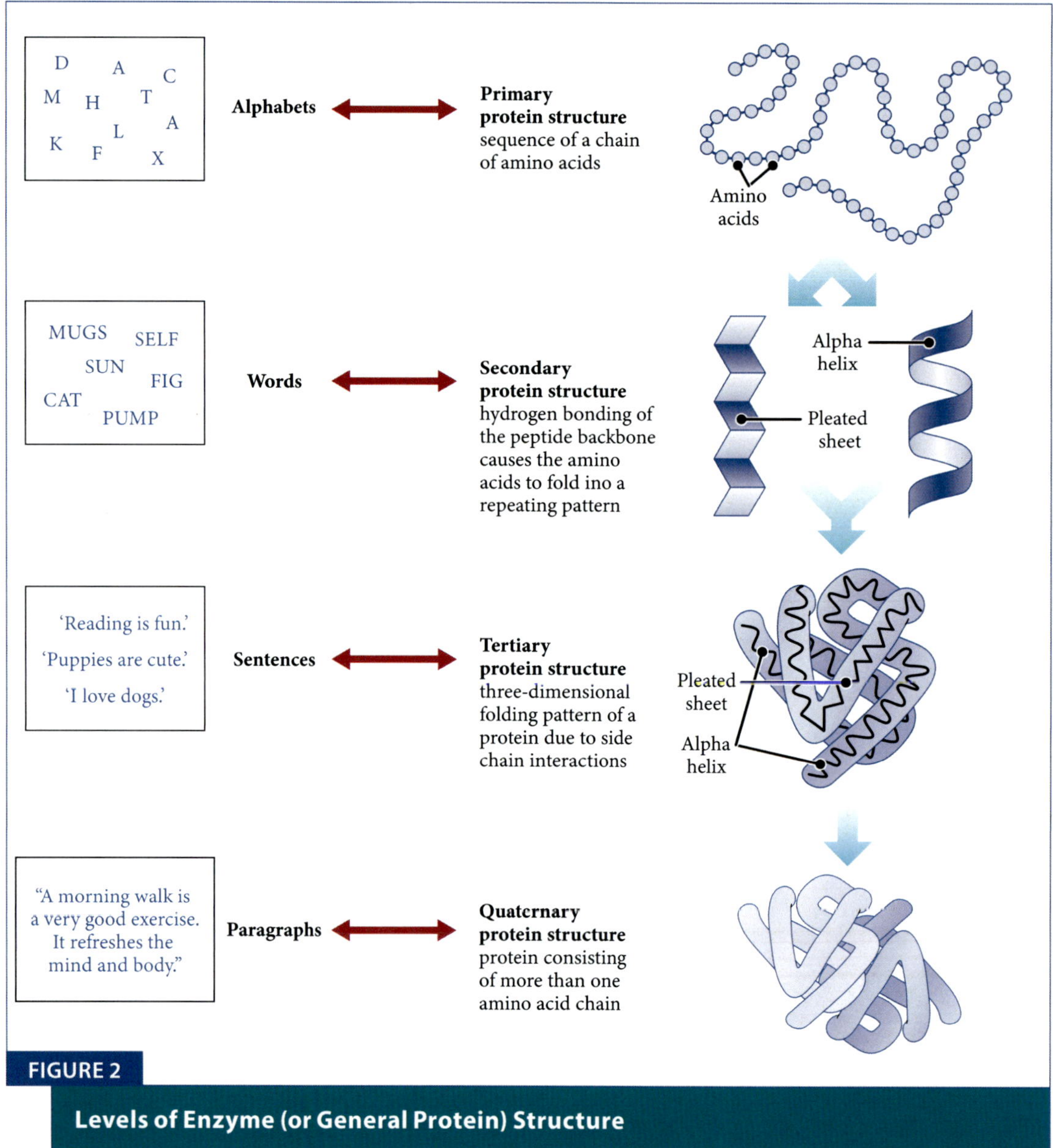

FIGURE 2

Levels of Enzyme (or General Protein) Structure

4. Incubate a sample of each bacterium at each temperature for 24 hrs.

5. After incubation, vortex each sample and use a pipetter (with tip) to subsample each into a **cuvette.** Fill the cuvette to the fill line to ensure that you have enough subsample to accurately measure it with the **spectrophotometer.**

6. Be sure that you use a TSB blank to properly calibrate your measurements.

7. Measure and record the results for each sample in one of the spectrophotometers. Use the data to construct a plot of absorbance or optical density vs. temperature.

ENVIRONMENTAL OSMOTIC PRESSURE

A very generalized definition of **osmotic pressure** is "the minimum pressure which needs to be applied to a solution to prevent the inward flow of its pure solvent across a semipermeable membrane" (Voet 2001). To make it particular to our situation, let us change a couple of key words so that we have, "the minimum pressure which needs to be applied to a solution to prevent the inward flow of [water] across [the cytoplasmic] membrane."

In this scheme, **osmosis** is the **net movement** of water molecules across the cytoplasmic membrane from an area of high concentration to low concentration. Concentration of what, again? Water. Water moves from where it is in high concentration to where it is in low concentration. "High" and "low" are relative terms that can only have meaning in relation to something. Higher than what? Lower than what?

One way to think about osmolarity is in terms of tonicity, which is a view from the perspective not of the water (the solvent) but of the solutes (dissolved substances, such as salts). From that perspective, **hypotonic solutions** are those that have a low concentration of solutes and a high concentration of water relative to the solution on the other side of the membrane. **Isotonic solutions** are those that have solute and water concentrations equal to that on the other side. **Hypertonic solutions** have a higher concentration of solutes and lower concentrations of water. So, in all these terms is the implied comparison between the external environment and the internal, cytoplasmic environment.

This **osmotic pressure of the environment** plays an essential role in the survival and reproduction of microorganisms. The cytoplasm of a generalized cell is ~80% water and ~20% dissolved and colloidal substances. For animal cells, which are bounded only by a fragile cellular membrane, a nearly isotonic environment is essential. The range of tolerable osmotic pressures is much wider across the microbial world. In all cases, though, a concentration of solutes and water can be found to induce **plasmolysis** which is the shriveling of cells as water leaves them or **plasmoptysis** which is the swelling of cells as water enters them (Figure 1). In animal cells, an acute plasmoptysis results in cell lysis, but in microbes such as bacteria, their cell walls provide a measure of rigidity that can often withstand low osmotic pressures. It is, in fact, common for bacteria to grow best in slightly hypotonic environments in which they maintain a slightly distended, turgid state. It should therefore become clear that various osmotic pressures are not uniformly helpful to all microbes and can therefore be exploited as an antimicrobial agent.

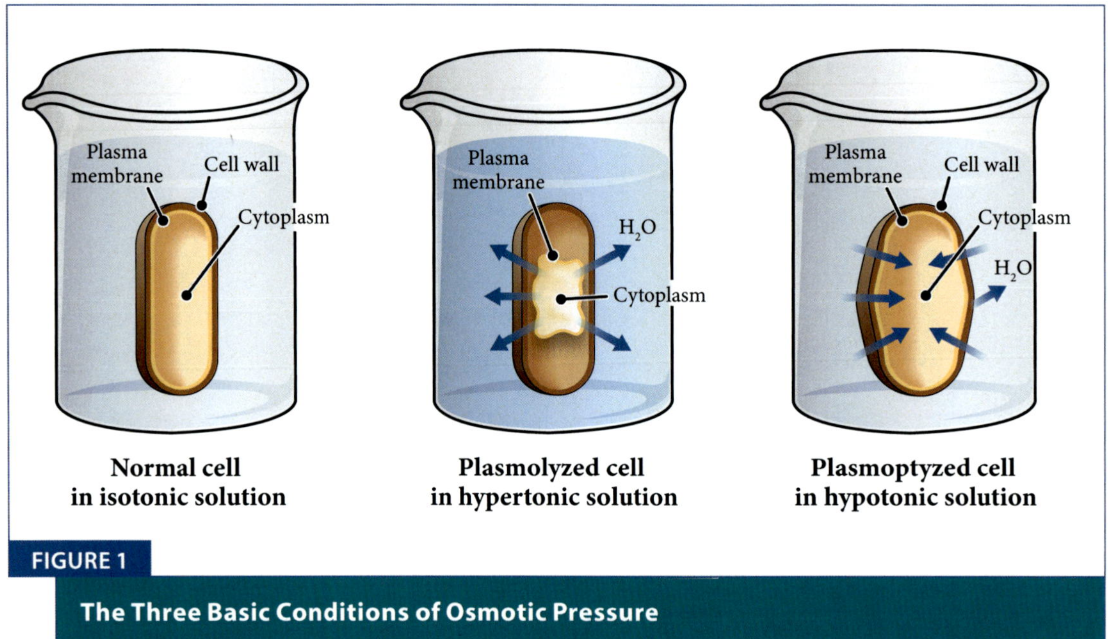

FIGURE 1

The Three Basic Conditions of Osmotic Pressure

Because of the reasons already listed, hypotonic solutions are not very effective as antimicrobial agents. Hypertonic solutions (high osmotic pressure) are fairly effective and are commonly used to inhibit microbial growth. Effective concentrations are not uniform because, as we have seen, microbes exhibit a tremendous diversity. If we use table salt (NaCl) as our model solute, we find that most microbes live and adjust to some range within 0.5% and 3.0% NaCl though there are plenty that can tolerate a bit higher. Concentrations of 10%–15% will completely inhibit the growth of most microbes, and those that are not completely inhibited are deemed **halophiles.**

The purpose of this laboratory exercise is to investigate the effects of different osmotic pressures on the growth of several bacterial species.

PROCEDURE

1. Obtain cultures of the relevant bacterial species and a TSB tube of each NaCl concentration.

2. The growth media is a standard TSB to which has been added 0%, 2%, 4%, 6%, and 10% NaCl.

3. Inoculate a tube of each NaCl concentration with each bacterium.

4. Incubate at 37° C for 24 hrs.

5. After incubation, vortex each culture tube and use the contents of each tube to fill a cuvette.

6. Measure the optical density of each cuvette using the spectrophotometer.

7. Plot the data for inclusion in the write-up.

REFERENCES

Voet, Voet, & Pratt (2001). Fundamentals of Biochemistry (Rev. ed.). New York: Wiley. p. 30. ISBN 978-0-471-41759-0.

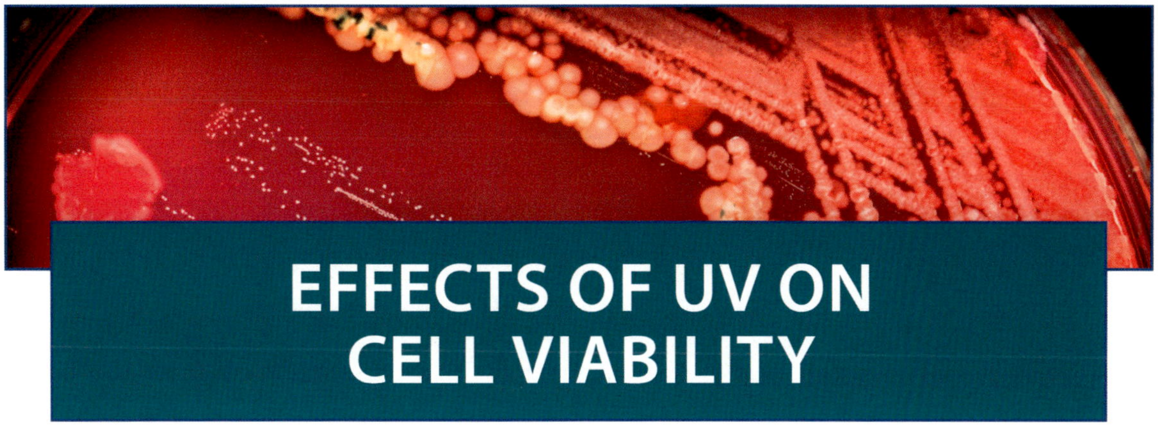

EFFECTS OF UV ON CELL VIABILITY

Electromagnetic waves (or are they particles?!) exist on a spectrum (Figure 1). There are two basic classes of radiation: **ionizing** and **non-ionizing** radiation. Both types consist (on the particle view) of **quanta** with different energies. Ionizing radiation includes things like X-rays and gamma rays which can knock electrons, protons, or neutrons out of molecules of the atoms they pass through. Non-ionizing radiation encompasses everything from ultraviolet light, visible light, infrared light, and up to radio waves. These non-ionizing **wavelengths** may still have certain effects on cells, however. In this laboratory exercise, we are focusing on the **ultraviolet (UV)** section of the spectrum.

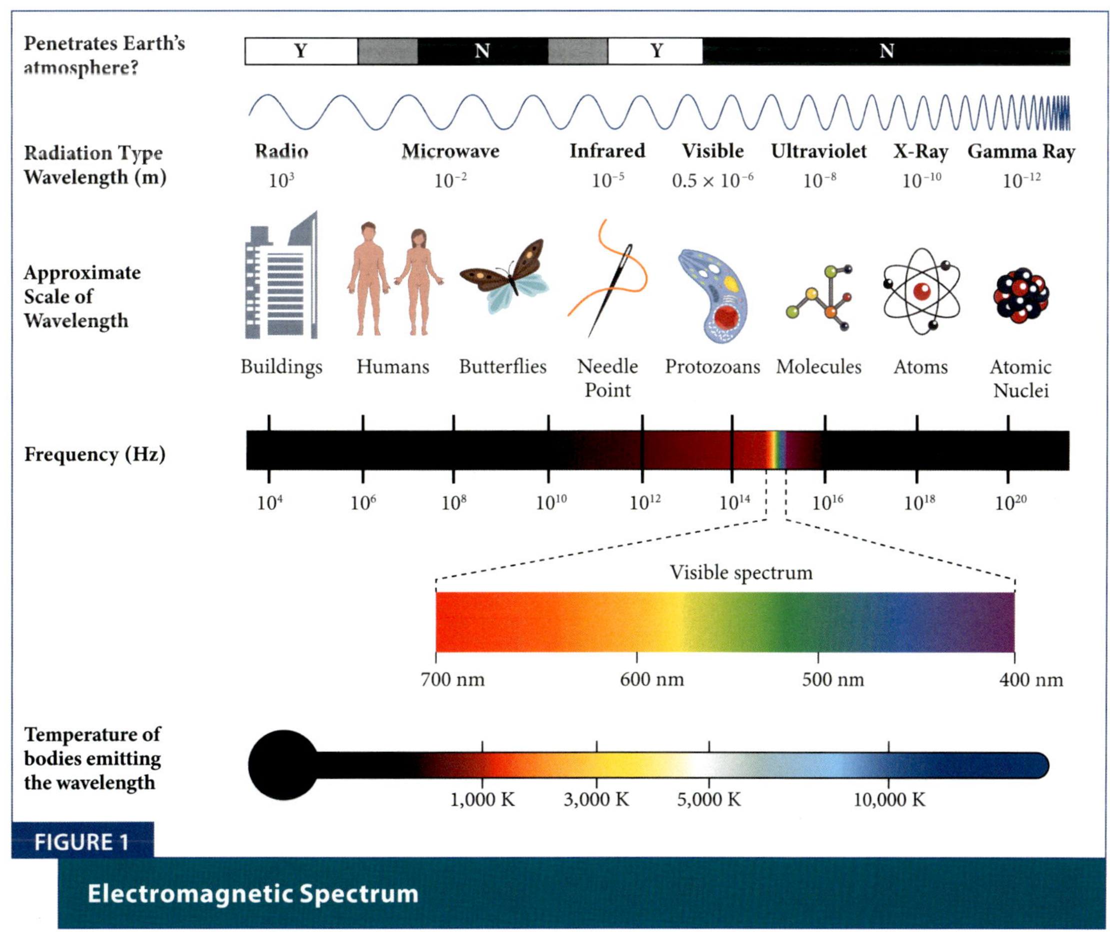

FIGURE 1

Electromagnetic Spectrum

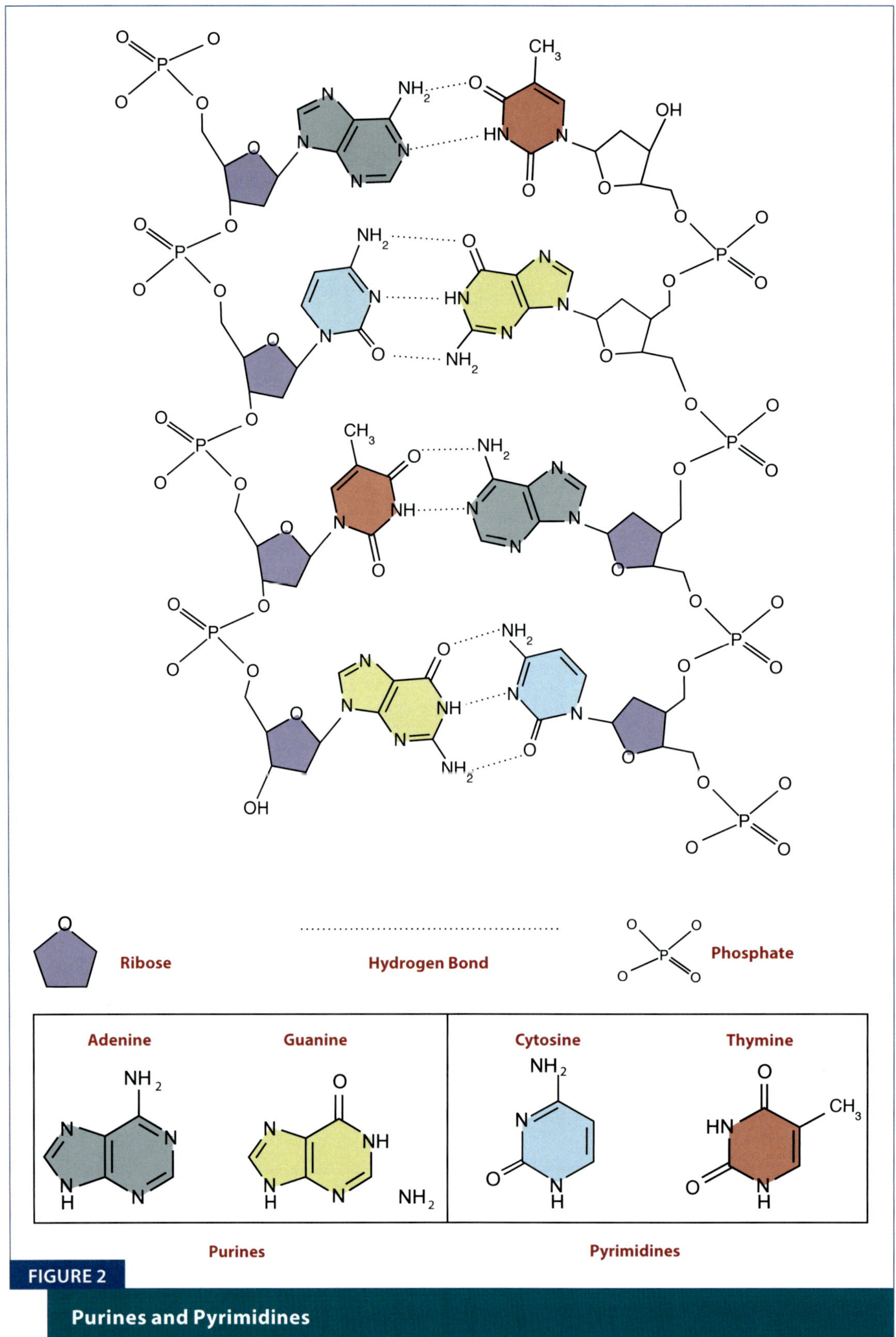

FIGURE 2

Purines and Pyrimidines

To understand the major impact that UV light may have on living cells, you first must understand the difference between the two basic types of nucleotide: **purines** and **pyrimidines.** Purines include adenine and guanine (A and G), and the pyrimidines include cytosine, thymine, and uracil (C, T, and U). They are classed differently because their nitrogen bases differ (Figure 2). Both the purines and the pyrimidines strongly absorb UV light (max for DNA and RNA is ~260 nm). Therefore, the major site of UV cell damage is the pyrimidine nucleic acid bases, due to their chemical structure.

The specific damage done by UV light is in the formation of **pyrimidine dimers** wherein two pyrimidines (C or T) adjacent to one another on the same strand of DNA become covalently bonded to one another. This can impede both the speed of DNA polymerase and its function by greatly increasing the likelihood that it will misread the sequence at the point of the dimer. The greater the density of UV quanta and the longer the exposure duration, the more dimers will be formed.

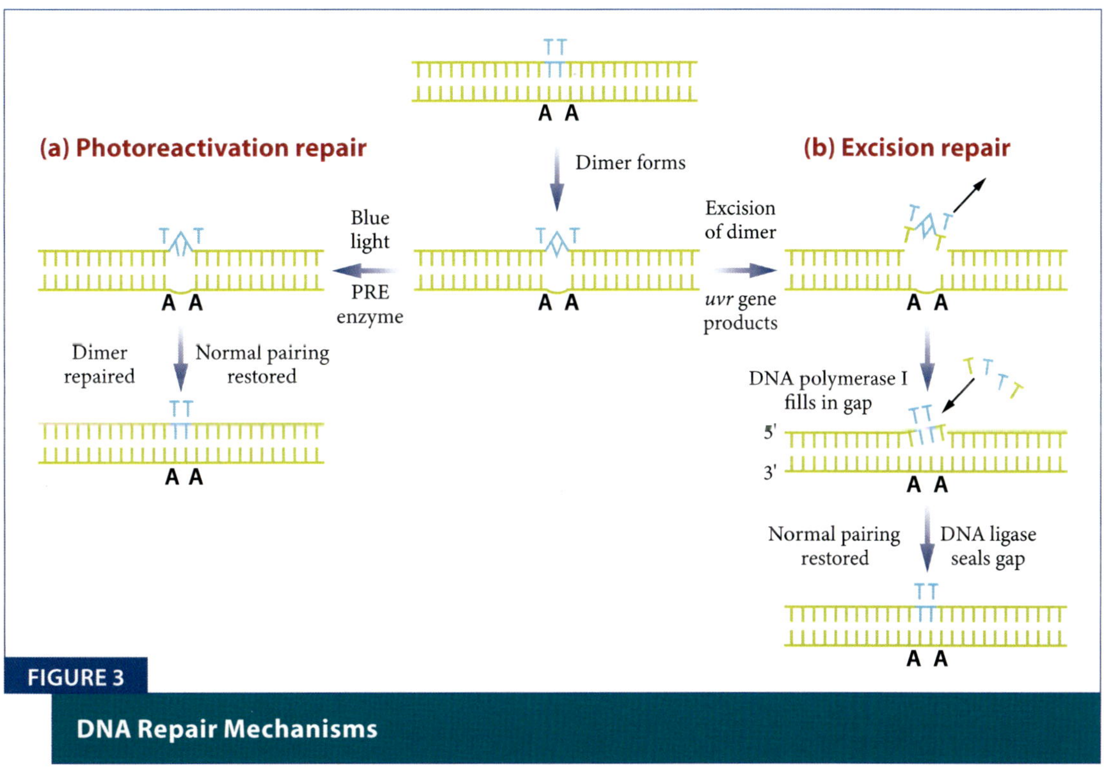

FIGURE 3

DNA Repair Mechanisms

Microbes have a number of different DNA repair mechanisms that they may use to fix synthesis errors or unlink dimers. Two basic types that can be active when a cell is exposed to UV light are **light repair** and **excision repair** (Figure 3). Light repair is activated when a cell has been exposed to light within range of 420–540 nm. This exposure activates a repair enzyme that will split any dimers that were formed. Excision repair is a bit more extreme and does not require light to activate it. Excision repair, as implied in the name, includes detection of a **DNA lesion** (error), **nicking** of the DNA strand containing the error, removal of the offending sequence (or nucleotide), and the writing of a patch using the complementary DNA strand.

In this laboratory exercise, you will be exposing inoculated (but not incubated) plates to UV light for various time intervals. This exposure may or may not impact the viability of the cells on the medium surface, and it is your job to describe what, if any, pattern you see. Finally, it is important to note that UV light has very poor penetrability, so even something as thin as paper or window glass will greatly reduce the amount of UV quanta reaching the cells beneath it. Its use should therefore be limited to the disinfection of **line-of-sight** surfaces and air.

PROCEDURE

1. Obtain a culture tube of each bacterial species and enough TSA plates to encompass the exposure times and lidded exposure.

2. Label the plates to indicate that one species will be on one half and the other species will be on the other half.

3. Inoculate each side of each plate with the given organism using a simple straight-line inoculation.

4. ***Do not parafilm all the plates.***

5. Parafilm only the 0 second and lidded exposure plates.

6. Expose the plates in the UV box, lids off, for 15 seconds, 30 seconds, 1 minute, and 4 minutes. The lidded exposure will have its lid on and should be exposed during the 4-minute interval.

7. For the health of your eyes and skin, minimize your exposure to the UV light as you are working.

8. Incubate the plates at 37° C for 24 hrs.

9. After incubation, observe your plates and note:

    a. Whether the species' responses were the same.

    b. What was the effect of increased exposure duration?

    c. Did the presence of a thin plastic lid have any effect?

    d. Feel free to include a sketch or picture of your results in your write-up.

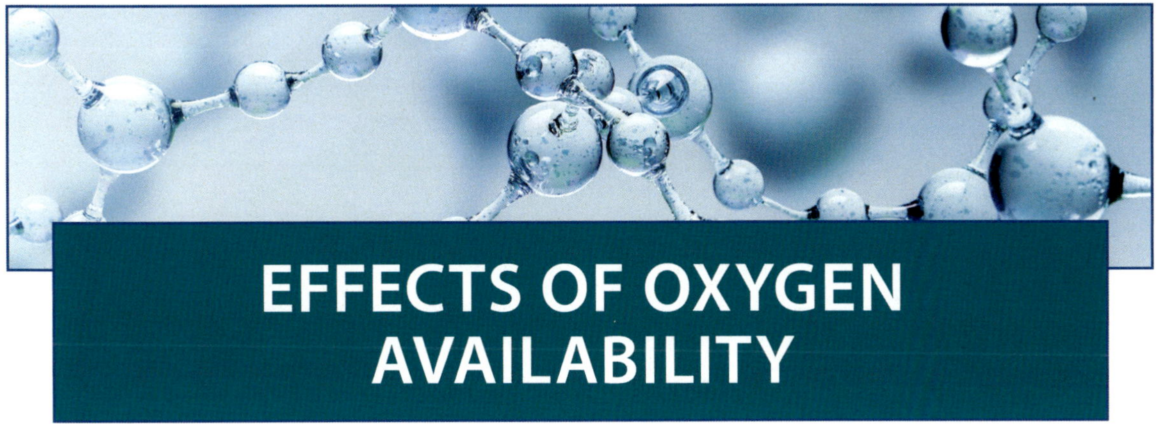

Microorganisms can survive and reproduce over a very wide range of **free/molecular oxygen (O$_2$)** concentrations. Much like we have discussed when we investigated the effects of temperature and pH, we can use the oxygen requirements microorganisms to classify them in a meaningful way.

A microorganism's oxygen (O$_2$) requirement is determined by that organism's compliment of bio-oxidative enzymes. This is what gives us a basis for five basic classifications (Figure 1):

- **Aerobes** – Require the presence of atmospheric concentrations of oxygen (O$_2$) for them to grow. They require it because their respiration systems use O$_2$ as their final e$^-$ accepter.

- **Microaerophiles** – Require O$_2$ but at concentrations lower than those in the typical atmosphere. If they are exposed to environments with atmospheric O$_2$ concentrations, the excess O$_2$ blocks the activity of their oxidative enzymes, killing the organism.

- **Facultative** – Also called "facultative anaerobes." Can be thought of as a subset of aerobes that are capable of respiring in O$_2$-low or O$_2$-free environments. These organisms will typically use other inorganic ions such as sulfate (SO$_4^{2-}$) or nitrate (NO$_3^-$) if there is no O$_2$ present, or they may use glycolysis + a fermentation pathway.

- **Anaerobes** – Also called "obligate anerobes," require the absence of O$_2$ because they do not express the enzymes to mitigate the harmful byproducts of aerobic respiration. If O$_2$ is present, the organism's biochemical machinery will use it as an e$^-$ acceptor and thereby create toxic byproducts such as H$_2$O$_2$ and O$_2^-$ with which the cells cannot cope. These organisms use other inorganic molecules as final e$^-$ acceptors.

- **Aerotolerant** – Here we have microorganisms that do not use O$_2$ as a final e$^-$ acceptor, and they are not poisoned by the presence of O$_2$. These organisms generally do not do respiration but rather employ only glycolysis + fermentation. They are also generally capable of producing **catalase** and **superoxide dismutase (SOD),** so the toxic compounds that kill anaerobes will not kill them.

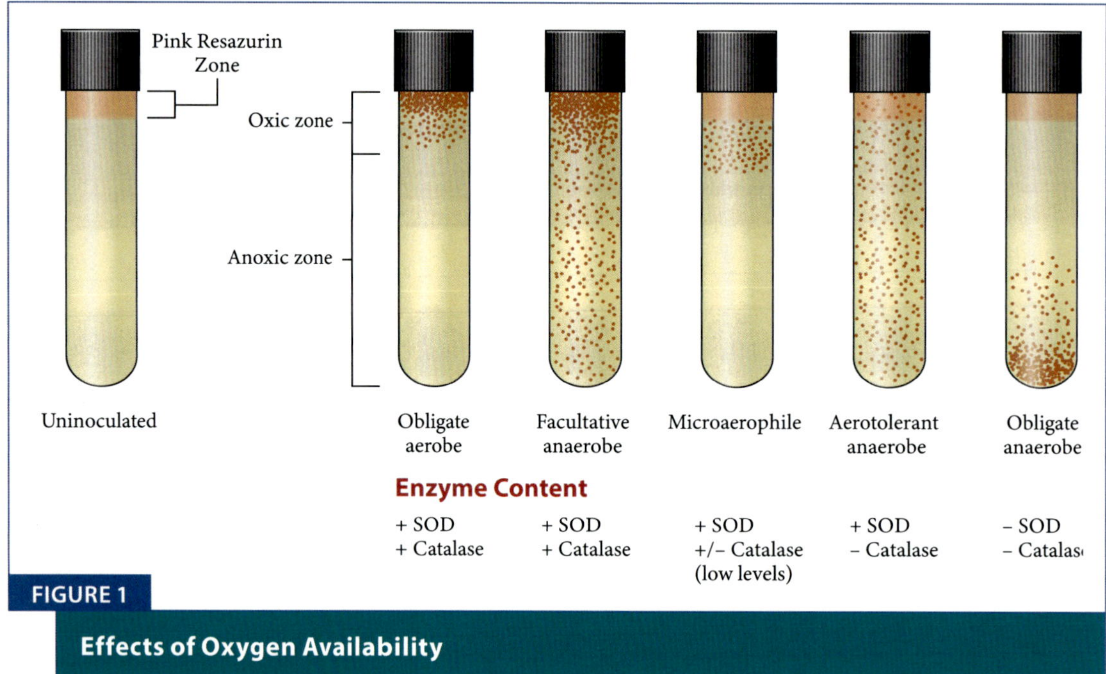

FIGURE 1

Effects of Oxygen Availability

In this laboratory exercise, you will investigate the oxygen requirements and abilities of a number of microorganisms. To effectively do that, you will need to use a medium that allows the formation of an **oxygen gradient.** The medium that you will use is fluid (semi-solid) thioglycollate (FTG) in which the sodium thioglycollate, thioglycollic acid, and L-cystine reduce the free oxygen in the medium to water. **Autoclaving** drives most of free O_2 out of the medium then the chemical constituents and the semi-solid consistency keep it low, at least for a period of time. The medium also contains **resazurin** which is pink in the presence of O_2 and colorless where O_2 is absent. Finally, the medium is semi-solid, so if you are careful in how you inoculate it, you will be able to determine whether your organism is motile.

PROCEDURE

1. Obtain a tube of FTG for each organism you are investigating.

2. Inoculate each tube with a single organism using a standard stab inoculation. That is, don't swirl the inoculation loop or stir the medium. Leave the medium as undisturbed as possible by stabbing directly downward, not touching the bottom of the tube, then remove the loop directly out along the line of inoculation.

3. Incubate at 37° C for 24 hours.

4. After incubation, observe the tubes with an eye toward:

   a. Whether there is any pink hue anywhere in the tube.

   b. At what depth(s) did the organisms grow?

   c. Did the organism stay in the inoculation zone, or did any of them begin to migrate outward toward the tube wall?

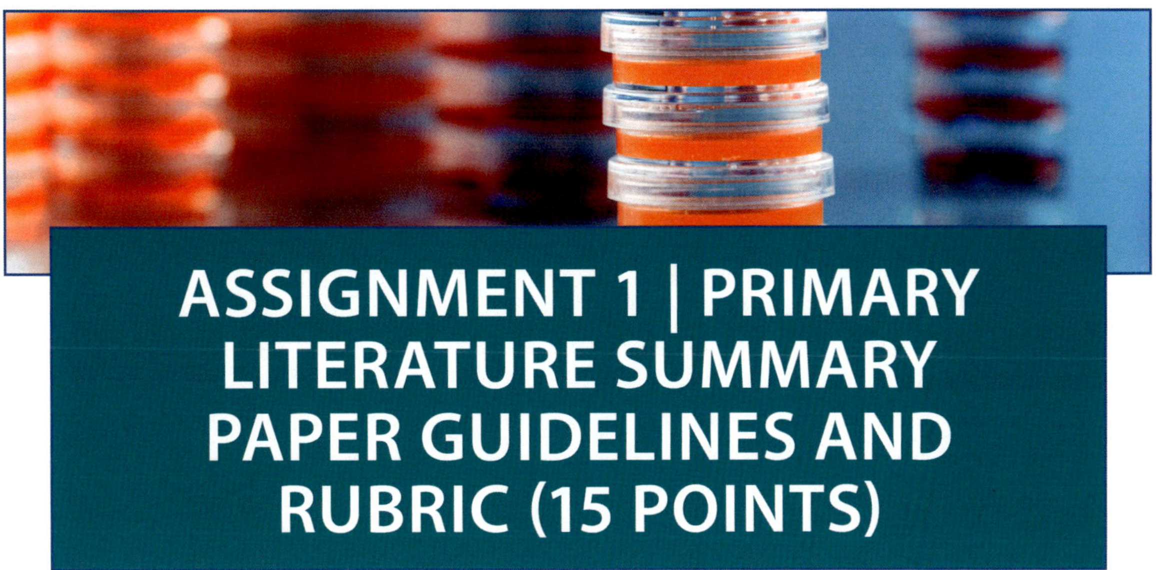

ASSIGNMENT 1 | PRIMARY LITERATURE SUMMARY PAPER GUIDELINES AND RUBRIC (15 POINTS)

Please read and summarize the posted paper titled "Strains, functions and dynamics in the expanded Human Microbiome Project." ***Electronic submission of this assignment will be due before the beginning of class.*** A hardcopy will be collected at the start of class as well. Failure to submit your paper online (**must be submitted to SafeAssign on BlackBoard)** will result in a zero for the assignment. Failure to hand in a hardcopy at the start of class will result in a 5% deduction. Each day the hardcopy is late will result in a 5% deduction.

GUIDELINES

- 1 page maximum (additional content will not be reviewed)
- Arial or Times New Roman font
- 11–12 pt. font
- Single-spaced
- Do not regurgitate the abstract
- No reference page and no in-text citation necessary
- Name and section number in the top left corner
- The electronic and hardcopy must be identical.

THINGS TO CONSIDER

- Define the purpose of the research outlined in the paper
- What are the major findings in the paper?
- What methods were used to gain the data?
- Overall significance of the results/findings?

- Use your words wisely. There is a 1-page maximum.

- Do not write an exhaustive summary of the introduction—a few sentences introducing the topic and stating the purpose of the research is enough.

- I recommend reading the paper multiple times and take notes throughout your readings.

- Make sure to organize the paper logically. Introduce the topic/purpose and then discuss the results in a small and broad context. You do not need to regurgitate each assay used in the paper; however, to explain the results, you will need to touch on the methodology. For example, do not list stepwise details of assays performed in the paper. Instead, provide something like the following: "The researchers employed homology-based WGS scans to determine homologous genes among the bacterial genomes."

RUBRIC (15 POINTS TOTAL)

CRITERIA	EXCELLENT (3 POINTS)	SATISFACTORY (2 POINTS)	UNACCEPTABLE (< 2 POINTS)
Summarizes the content of the article (S)	Includes a thorough, clear, and concise summary of the article.	Summary shows some evidence that the paper has been read but is not clearly summarized.	Summary is not a clear illustration of the article.
Includes appropriate level of detail (I)	Includes relevant details that aid the overall understanding of the article.	Includes *some* details that contribute to understanding the article.	Does not include any details that contribute to the understanding of the paper.
Overall organization (O)	Overall purpose, methods, results, and conclusions of study are clearly paraphrased in a logical flow.	All components included but the paper lacks a logical flow.	Major sections are missing.
Clarity (C)	Summary is comprehensive and competently uses scientific terminology.	Comprehensive but mostly in layman's terms.	Vague/general coverage lacking a clear view of the work.
Mechanics (M)	Free of grammatical errors and exhibits an easy to read writing style.	Some grammatical errors present or writing style is choppy.	Serious grammatical errors, constant misspelling, and writing is unintelligible.

1. **Please remove all labels** from your tubes when you are finished completing your tests and place them in the metal bins near the back of the room.

2. Remember to think about these tests/results in the context of *evolution and genetic diversity.*

3. *The final exam/practical will rely heavily on information surrounding these tests.*

4. The questions presented below are by no means comprehensive and are only present to prompt thinking.

GELATIN HYDROLYSIS

- Observe your gelatin stab agar tube for liquefaction.

- Is the gelatin agar solid or liquid?

- What enzyme is being tested for here?

- Is gelatin a complete protein? If not, what essential amino acid is missing?

- Is collagen derived from gelatin, or is gelatin derived from collagen?

GLUCOSE/SUCROSE FERMENTATION

- Observe your glucose and sucrose tubes for any color change.

- What does no color change indicate? What does a color change from red to yellow indicate? What end-products are being measured here? What is the name of the pH indicator?

- Observe the inverted tube (Durham tube) located inside each tube for gas production.

- How can you tell if gas has been produced or not?

- How many ATP are produced if glucose/sucrose fermentation is carried out?

METHYL RED (MR)

- *Before* adding Methyl Red to this tube, **transfer half of the contents into a sterile test** tube. This new tube will be used to conduct the Voges-Proskauer (VP) test (see below).

- Add 3–5 drops of the Methyl Red reagent to the MR tube and observe any color change.

- What does it mean if the broth remains clear/yellow? What does it mean if the broth turns red?

- What is being tested here? What is the pH indicator?

VOGES-PROSKAUER (VP)

- Add 5 drops of each Barritt's reagent (Barritt's A and B) to the broth.

- **Observe any color change after 15–20 minutes.**

- What does a pink/red coloration at the top of the broth indicate?

- What docs this test chemically identify?

- What is the name of the Barritt's A reagent? What is the name of Barritt's B reagent?

CITRATE UTILIZATION

- Observe your citrate agar slant for any color change.

- What does it mean if the media remains green?

- What does it mean if the media turns blue?

- What is the pH indicator?

- What enzyme is being tested for here?

- The production of _________________ combines with sodium and water, which forms _________________ and changes the _________________ indicator from green to blue.

NITRATE REDUCTION TEST

- Add 5 drops of Solution A followed by 5 drops of Solution B to the nitrate broth and observe any color change.
- What does a color change to red indicate?
- If the media does not turn red, then add a small amount of zinc powder to the tube and observe any color change. What is the purpose of adding zinc?
- What does a color change to red at this step indicate?
- What enzyme is being tested for?
- What is the name of Solution A?
- What is the name of Solution B?

SIM AGAR

- Add ~5 drop of the Kovac's reagent to your test tube.
- Does the media change colors?
- What does a color change (yellow→red) indicate?

FLUID THIOGLYCOLATE

- What growth pattern does your organism exhibit?
- Is there a pellicle, sediment, consistent growth throughout?
- What does this say about the oxygen requirements of your organism?